For Michael

Enjoy Exploring
Robert Piccioni

Everyone's Guide to

Atoms, Einstein,

and the

Universe

Real Science for Real People

Robert L. Piccioni, Ph.D.

Real Science Publishing
3949 Freshwind Circle
Westlake Village, CA 91361, USA

Visit our web site: **www.guidetothecosmos.com**

Einstein enjoying his sailboat and pipe

Cover and Interior Design by Fiona Raven: www.fionaraven.com
Edited by Arlene Prunkl: www.penultimateword.com
ISBN 13: 978-0-9822780-7-9
Second Edition, 2011
Printed in Canada

Another Award-Winning Book by Dr. Robert L. Piccioni
"Pinnacle Achievement Award Winner for Science"—NBEA

For decades, most scientists have believed that Life began when the right molecules randomly collided in just the right way—that "time itself performs the miracles." Other scientists have rejected such an accidental origin as being as unlikely as "the assemblage of a 747 by a tornado whirling through a junkyard."

Dr. Piccioni's remarkable second book examines the many prerequisites for Life—a universe with an exact geometry, the right atoms, a viable habitat, and an effective genetic code—and computes the likelihood that all these

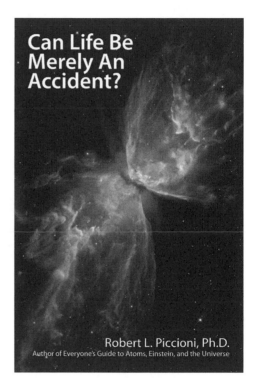

could arise by random chance. It demonstrates that our universe is tuned, with amazing precision, to support Life, and that our DNA is the most complex and precise structure in the entire cosmos. An explicit and comprehensive analysis concludes that Life arising by random chance, through any known process, is *extraordinarily unlikely*—less likely that drawing the ace of spades thousands of times in a row.

Contents

1. Once Over Lightly   10

Part 1 – The Micro-World

2. Got Atoms?   16
3. Einstein Who?   20
4. Einstein Settles Atomic Debate   28
5. Inside the Atom   34
6. Elementary Particles   38
7. Antimatter   46
8. Four Forces of Nature   50
9. Energy, Mass, and $E=mc^2$ 60
10. Smart Energy   66
11. Particles and Waves   74
12. Galileo and the Principle of Relativity   82
13. Einstein's Theory of Special Relativity   90
14. Einstein and Light   100
15. Al Makes Mom Proud   104
16. Einstein and Quantum Mechanics   108
17. Quantum Mechanics after Einstein   120

Part 2 – Stars

18. Twinkle, Twinkle Little Star…   134
19. Newton and Einstein on Gravity   144
20. Einstein's Theory of General Relativity   152
21. Solving Einstein's Field Equations   162
22. General Relativity in Action   166
23. White Dwarfs   170
24. Neutron Stars   176
25. What Are Black Holes?   182
26. The Care and Feeding of Black Holes   192
27. NASA's Great Observatories   198

6

Part 3 – The Universe

28. How Much? How Large? How Old?   204
29. What Is Our Universe?   218
30. Telescopes Are Time Machines   222
31. It's the Same Everywhere   226
32. Redshift / Blueshift   230
33. Expansion   234
34. CMB: the First Light   246
35. Dark Matter   254
36. Dark Energy   260
37. Our Special Place in the Cosmos   264
38. Can We Save Earth?   270
39. The Big Bang   274
40. What Came Before?   288

Color Plates: 1 – 33

Bibliography   294
Glossary of Terms   298
Index   308
Table of Redshift vs. Time   317
List of Symbols   318
More Einstein Quotes   320

List of Sidebars

Refugees Help U.S. Fission while Fascists Fizzle   45
Why Fund Science?   81
No Nobel for Relativity   103
Fermi's Delicious Book   225
Einstein's Expensive Divorce   245
Silk Stockings Save Dad   253
Feynman at Los Alamos   259
Feynman at Caltech   259
Shooting Pool with Feynman   269

Dedicated to Joan

For saying, "You should write a book."

For patiently listening to all my lectures and critiquing countless drafts. Joan made this a much better book.

But more than anything else, for being my partner and my inspiration for 41 years. Together we rejoice in life, wonderful children, and delightful grandchildren.

SUNRISE AT MACHU PICCHU, PERU, 2007

Acknowledgments

I applaud the organization and the people of NASA and its affiliates for their genius, diligence, perseverance, sacrifice, and achievement in immeasurably advancing science and freely providing all their discoveries to the whole world. The lion's share of all the spectacular celestial images in this book come from NASA's incomparable space telescopes— Hubble, Chandra, Spitzer, and WMAP.

I happily thank all the kind souls who provided invaluable assistance in creating this book.

Bill Sampson meticulously scrutinized an early draft and vigorously helped beat it into shape.

Gray Jennings, Gene Burke, and Frank Gari kindly read the "final draft" and provided cogent ideas.

Arlene Prunkl edited the text and put me in touch with the right professionals to get the project completed.

Fiona Raven created both the cover design and the interior design. She patiently tutored me in the many facets of creating a book. Fiona was delightful to work with and an indispensable guide.

My brother Richard Piccioni, Ph.D., a high school physics teacher, introduced me to the joy of speaking to a future generation of scientists, and provided many valuable suggestions.

Most of all, I thank my wonderful wife Joan, who drained the Red Sea marking up countless, very rough drafts, and who kept our lives afloat while I obsessed over every detail. The credit for making the explanations of complex topics understandable belongs to Joan, who never let me get away with "almost good enough."

1

Once Over Lightly

You don't need to be a great musician to appreciate great music. Nor do you need great math or physics expertise to appreciate the exciting discoveries and intriguing mysteries of our universe.

We live in the Golden Age of Science. More has been discovered in the last century than in all prior history. We are the first generation to have seen almost to the edge of our universe and almost to the beginning of time. We have seen the largest structures that exist in nature, and have discovered its smallest parts. For the first time, physical science has a coherent story of almost everything.

Our adventure begins by diving into the infinitesimal, then soaring through the stars, and ultimately reaching for the limits of the universe. We will discuss almost all the major discoveries of modern physics, astronomy, and cosmology, and encounter several recurring themes along the way:

- Everything is intimately connected, from the smallest to the largest, from atoms to people to the universe.
- Things that seem vastly different are often really the same.
- Einstein's belief in the simplicity and beauty of nature inspired him to find unity and elegance in science.
- Great minds with extraordinary vision often achieve great strides— but sometimes can't see small steps that are so clear to others.

Part 1 examines our long journey to discover the *micro-world*—molecules, atoms, and everything smaller—and introduces those who led the way.

For 2500 years, some of the brightest minds struggled to discover what everything we see is made of. In 1905, Einstein provided the critical insights that firmly established that matter is made of atoms. But atoms are not the end of this quest. Atoms are composed of *electrons* and *nuclei*. Nuclei are made of *protons* and *neutrons* that are themselves made of *quarks*. Finally, we think we have reached the innermost layer of matter—twelve *elementary particles*. We have also discovered an unexpected bonus: *antimatter.*

The interactions of these elementary particles underlie everything in the universe through nature's four forces: *gravity, electromagnetism, strong,* and *weak.* These forces light up the stars, create the atoms in our bodies, enable all chemical and biological processes, preserve Earth's atmosphere, and shape planets, galaxies, and the universe itself.

We will learn how an obscure clerk, rejected by the academic establishment, single-handedly shook the foundations of science and forever changed our understanding of *energy, mass, light, space*, and *time.*

Einstein's most famous equation $E=mc^2$ provides a deeper understanding of mass and energy that can lead us to develop dramatically more abundant and less polluting sources of energy. Future energy production can be a million times more efficient and less polluting than current technology. Ultimately, by utilizing *black holes*, we may be able to provide all the energy needs of a million people for 1¢ per day, and do so with zero pollution.

Einstein's two theories of Relativity are among the crowning achievements of 20TH century science. Building on the discoveries of Galileo, and extending the scope of Newton's laws, Einstein opened the universe to science. He said that different clocks, even perfect clocks, keep time differently depending on their speed and location. There is not just one right answer to the question, "What time is it?" Einstein explained why time is *relative*, and why a jet looks shorter and heavier the faster it flies. Einstein's theories constrain and enable distant space travel and also raise puzzling questions, such as the twin paradox.

We will examine the mysteries of Quantum Mechanics and its startling view of reality in the micro-world. We'll also learn how Quantum Mechanics made possible the electronic revolution that permeates our lives through computers, cell phones, and all things digital. We will meet Schroedinger's Cat and discover who had the last meow. Einstein made several essential contributions to the development of Quantum Mechanics, including establishing the theoretical basis for the lasers that scan our bar codes, read our CDs, print our documents, and sculpt our corneas. His contributions ultimately led to a quantum view of reality filled with uncertainty. However, Einstein himself never accepted the uncertainty of Quantum Mechanics, declaring "God does not play with dice."

Part 2 explores how the *micro-world* impacts the *macro-world* (everything larger than molecules). In particular, the properties of particles control the stars that have transformed the cosmos from a cold, empty, lifeless void to a universe of spectacular sights and endless possibilities. Stars both enable life and frame our future.

We will follow the life cycle of stars: their birth, the exquisite balance of their prime, their rapid decline, and their ultimate deaths in immensely violent explosions called *supernovae*. Through these explosive deaths, nature recycles vital resources, plants the seeds of rebirth, and leaves exotic remnants such as *white dwarfs, neutron stars*, and *black holes*. With spectacular celestial photographs, we will examine the bizarre nature of each of these exotic remnants, particularly the most enigmatic: black holes.

Einstein's Theory of General Relativity enables the Global Positioning System (GPS). It also provides the foundation for our understanding of stars, galaxies, and the universe. General Relativity is widely accepted as the most beautiful theory in physics, and many believe it is the greatest achievement of human thought. Can a theory really be considered beautiful?

We will examine NASA's wonderful space telescopes, particularly the Hubble Space Telescope, the greatest advance in astronomy since Galileo first pointed a telescope toward the heavens 400 years ago. NASA's space telescopes have opened up the heavens as never before.

Part 3 builds on our knowledge of both the micro-world and the macro-world of stars to illuminate the mysteries of the universe. We begin with an exploration of our universe as it is today. How big are galaxies? How many galaxies are in the universe? How large is the universe? How small is its smallest part? Where do we fit into all this? What do we know, what don't we know, and what might we never know?

Next, we turn to cosmology and explore how the universe, and our understanding of it, has evolved. Cosmology became a quantitative science in the 20TH century with Einstein's Theory of General Relativity, the observations of Henrietta Leavitt and Edwin Hubble, and the meticulous study of starlight.

Everything astronomers know about the cosmos comes from observing starlight. It's amazing how much we can learn from the charming twinkle of stars. From the array of "colors" in starlight, astronomers can precisely measure what stars are made of. From changes in these color patterns—*redshifts* and *blueshifts*—they measure the motions of stars and the universe.

Leavitt discovered the key to measuring the distances to very remote stars. Hubble built on Leavitt's discovery to demonstrate that there are galaxies far beyond our own Milky Way. Then, Hubble used redshifts, as well as distances derived from Leavitt's technique, to discover that the universe is expanding. We will learn the meaning of this expansion, what is expanding, and what is not.

Some discoveries are the result of many years of careful preparation and precise observation. Others are fortuitous accidents, such as the detection of the afterglow of the Big Bang. This accidental discovery, followed by decades of meticulous measurement, provides a grand story about the very early universe.

Further, we will examine the special and wonderful place mankind occupies in this vast universe, in a most favorable place, at a most favorable time, and enjoying a most fertile habitat.

Next, we turn to cosmology's greatest achievement, the Big Bang Theory, starting with the beginning of space and time. We will examine the critical role played by the dark side of our universe: *dark matter* and *dark energy*.

Finally, we will explore the most promising ideas about what came before the beginning and what lies beyond. While the rest of this book is based on well-established science, this last discussion involves intriguing speculations.

Our quest is similar to a grand buffet. Feel free to pick and choose. If you don't care for anchovies, don't fret, just skip on to the next delight. If you don't enjoy Quantum Mechanics, simply move on to Stars. To make these cutting-edge concepts as accessible as possible, I translate physics into English, replace equations by graphics, and provide a gallery of heavenly pictures.

But the science is not "dumbed down."

This is real science for real people, like you.

You will have much to think about.

For your convenience, a Glossary of Terms and a List of Symbols are provided at the end of this book. Scientific terms and words that have a special meaning in science are *italicized* when first introduced. You will sometimes see bracketed numbers such as [1] that indicate further discussion is in notes at the end of that chapter.

I hope you enjoy and benefit from reading this book. If you have any comments, suggestions, or questions, please contact me at:
www.guidetothecosmos.com.

NOTES

[1] For those not allergic to math, some more technical matters are explored in notes at the end of chapters.

PART 1

The Micro-World

Atoms

   Particles

      Forces

         Energy

            Relativity

               Quantum Mechanics

2

Got Atoms?

For over 2500 years, people wondered:

"What is everything really made of?"

Now we think we know.

The material in everything we see seems smooth and *continuous*, like the paper in this book. When we pour a glass of water, we don't see pieces of water fall out; we see a continuous stream. For centuries, many people believed matter remained continuous, even down to the very smallest possible dimensions. But others believed that matter was ultimately made of separate and *discrete* parts that were too small to be seen.

Consider a simple one-dimensional example: a long, thin wire. If matter is continuous, we could examine the wire under a microscope, as in figure 2.1, and even at extreme magnification it would still look smooth and continuous. It seems wire can be cut to any desired length no matter how small, and even the smallest piece retains the essence of being wire.

But if matter is ultimately made of separate and discrete parts, at high enough magnification, a wire might look like a chain made of individual links. Chains can have only certain lengths; they can be 2 links long, or 200 links long, but never 2½ links long. And unlike a wire, if a chain is

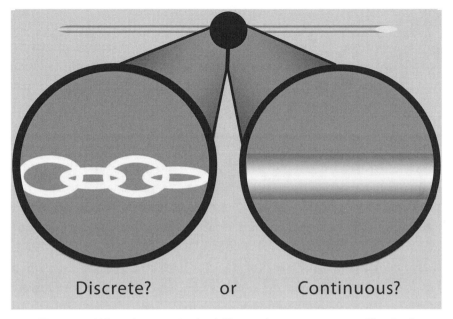

Figure 2.1. What does a wire look like under extreme magnification?

cut into ever smaller pieces, eventually the pieces are no longer chain-like. The essential character of being a chain is lost once the links are cut.

Democritus, a Greek philosopher, mathematician, and astronomer, is credited with being the first to name the small, discrete pieces from which he believed all material objects are made. He called them *atomos*, Greek for *uncuttable*.

The "atomic" debate continued for millennia as an entirely philosophical discussion. No one could conclusively prove or disprove the existence of atoms.

In the 19TH century, physicists developed a theory of heat, called Thermodynamics, which is very effective in describing the behavior of gases and in designing efficient steam engines. Thermodynamics deals with aggregate properties of matter, such as volume, pressure, and temperature. Its laws have remained virtually unchanged for over a century and are still used to design all heating and cooling systems.

In the second half of the 19TH century, Ludwig Boltzmann, an Austrian mathematical physicist, searched for a deeper understanding of heat. He wanted to learn why the laws of Thermodynamics work so well.

What causes a gas to have a temperature and a pressure, and why do these aggregate properties of matter behave as the theory says? Using Newton's laws of mechanics and the mathematics of probability and statistics, Boltzmann developed a successful theory based on the assumption that matter was made of a vast number of very small atoms.

Boltzmann's theory is called Statistical Mechanics. It explains why matter behaves according to the laws of Thermodynamics. His theory says that temperature is due to the vibration of atoms and that atoms in a hot object move faster and have more energy than atoms in a cold object. The theory explains why faster atoms transfer energy to slower atoms until their energies and temperatures equalize. This explains our everyday experience that an ice cube melts in hot tea and cools the tea. It also explains why pressure increases at higher temperatures: when a bottle of gas is heated, the gas atoms get hotter, move faster, and hit the bottle's walls more often and with more force.

Despite the successes of Boltzmann's theory many scientists challenged him and vehemently attacked his atomic assumption as unproven and unjustified. They demanded direct evidence: "If atoms exist, show me one." In particular, Austrian physicist Ernst Mach (known for the *Mach number* for measuring airspeed) and German physicist Wilhelm Ostwald promoted an alternative theory called *energetics* that rejected the existence of atoms. Mach and Ostwald were more effective than Boltzmann in swaying scientific opinion. In the absence of definitive evidence, enthusiasm for the atomic theory waned.

Unfortunately, Boltzmann had a long history of rapid and severe mood swings. In failing health and despondent at what he perceived to be the rejection of his life's work, he committed suicide.

A few years later, Boltzmann's belief in atoms was vindicated, thanks to an obscure clerk in the Swiss patent office.

A century later, we can answer Boltzmann's critics even more directly. No longer do we need to just imagine atoms, now we can actually image *individual* atoms, as shown in figure 2.2.

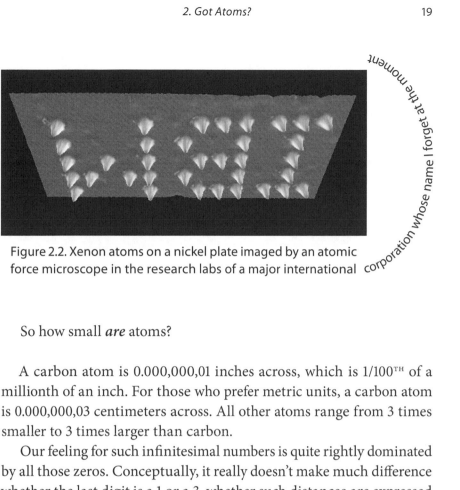

Figure 2.2. Xenon atoms on a nickel plate imaged by an atomic force microscope in the research labs of a major international corporation whose name I forget at the moment.

So how small *are* atoms?

A carbon atom is 0.000,000,01 inches across, which is 1/100ᵀᴴ of a millionth of an inch. For those who prefer metric units, a carbon atom is 0.000,000,03 centimeters across. All other atoms range from 3 times smaller to 3 times larger than carbon.

Our feeling for such infinitesimal numbers is quite rightly dominated by all those zeros. Conceptually, it really doesn't make much difference whether the last digit is a 1 or a 3, whether such distances are expressed in inches or centimeters. As you continue through this book, you will encounter very small numbers and very large numbers. Don't bother memorizing numbers or fretting about English versus metric units— there's no test at the end. Instead, try to get a feeling for the scale of the numbers and focus on the concepts.

3

Einstein Who?

Albert Einstein was born in 1879, in the city of Ulm in southwestern Germany. Figure 3.1 shows a photograph of young Albert. His parents were non-observant Jews whose ancestors may have lived in that part of Europe for two centuries. They considered themselves to be more German than Jewish. When the Nazis came to power, that would change. [1]

YOUNG ALBERT

Einstein did not have an auspicious start in life. In fact, he was 30 years old before he first began to achieve success in his lifelong ambition: to become an academic scientist, and in particular, a theoretical physicist.

As a child, Einstein started speaking late, and even then his speech was often awkward. Members of the household called him "the dopey one." His sister Maja, two years younger and his only sibling, said Albert rehearsed everything he said, slowly moving his lips, before speaking aloud. Nevertheless, eventually it became clear that young Albert was indeed very bright. Perhaps his poor communication skills were due not to a lack of intelligence but to a lack of interest in interacting with others. Throughout his life, Einstein's interests were focused internally. He was intensely absorbed with his own thoughts and his fascination with the beauty of nature. His internal focus was so intense that he could often "tune out" the noise and chaos of the outside world and concentrate entirely on his own ideas.

Above all else, Einstein prized personal freedom. He was insolent, defiant of authority, and vigorously contested all attempts to force him into conformity. Einstein proudly declared himself a "lone wolf", determined to live life on his own terms, with little concern for the opinions of others. He rebelliously declared: "Long live impudence" and "Blind respect for authority is the greatest enemy of truth."

Figure 3.1. A young Einstein

Einstein despised the highly regimented German schools of the day, and his nonconformist spirit did not sit well with school authorities. Young Einstein was expelled from one school and asked to leave another. The headmaster sternly told Hermann Einstein, Albert's father, that his son would never amount to anything. As it turns out, the headmaster was almost right.

When Einstein was 15 years old, his father's business suffered one of several reversals, and the family moved to Italy for better prospects. Albert was left behind with relatives to finish the last three years of high school in Germany. That didn't last long. The following autumn, Einstein dropped out of high school. Some believe the school administration forced him to leave. In any case, the authorities had no regrets at Einstein's departure. He rejoined his family in Italy and renounced his German citizenship. He may have been motivated in part by Germany's compulsory military service for all 17-year-old males that was anathema to Einstein.

He then applied for admission to the Polytechnic Institute in Zurich, Switzerland, which was considered a science and math teacher's college.

It was not in the top tier of European universities, and had no Ph.D. programs. However, Einstein failed the entrance exam, with particularly poor marks in French.

Advised to finish high school and then reapply, Einstein enrolled in a school in the nearby Swiss village of Aarau. This school's approach was based on the educational philosophy of Johann Pestalozzi, and was completely different from what Einstein had experienced in Germany. To his delight, they emphasized independent thinking, visualization, and nurturing each student's "inner child." Einstein hardly needed encouragement in independent thinking, but the emphasis on visualization may have had a very positive and lasting influence.

Einstein said that he rarely thought in words: "A thought comes, and I may try to express it in words afterwards." It was in Aarau that Einstein had the first thought that ultimately led to his Theory of Special Relativity. He tried to imagine what he would see if he could run alongside a light beam at the speed of light. He reasoned that he should see the light beam standing still, but how could a light wave exist without motion? This was a dilemma that took Einstein a decade to resolve [2].

COLLEGE YEARS

The next year, he reapplied to the Polytechnic and was admitted. Initially, all went well for him. At the end of his second year, he ranked first in his class. But gradually, Einstein's insolence and disrespect of authority became issues again, and he soon earned the displeasure of his professors. Hermann Minkowski, his mathematics professor, said he was a "lazy dog." Einstein habitually called one physics professor "Herr Weber" instead of the proper honorific, "Herr Professor Weber"—a deliberate and seemingly pointless affront. Einstein skipped classes he found boring and chose not to complete assignments he considered beneath his intellectual stature. Another physics professor, Jean Pernet (not the patriarch of a famous champagne label), has the unique distinction of giving Albert Einstein a failing grade in a physics course, Physical Experiments for Beginners. As we will discuss later, the only experiments Einstein liked to perform were those he could do in his own mind: *thought experiments.* Pernet was so displeased that he convinced the director of the Polytechnic to officially reprimand Einstein for his lack of diligence.

In 1900, after the customary four years, Einstein graduated from the Polytechnic in the lower half of his class. With low marks and strongly negative references, particularly from "Herr Weber", doors closed in his face all across Europe. For years, Einstein applied for jobs at every university and technical institute from Norway to Italy and was summarily rejected. He was the only member of his graduating class not to secure professional employment. He tried to eke out a living through part-time tutoring, but was forced to fall back on the financial support of relatives.

MALEVA MARIC

While at the Polytechnic, Einstein met Maleva Maric, the only female student in his class. Maleva was Serbian, three years older than Albert, and very strong in physics and math—in fact, she was better in math than he. Maleva is generally credited with providing key assistance with the mathematics and background research for some of Einstein's important early papers.

Albert and Maleva shared a passion for science and a contempt for the established conventions of society and academia. They became colleagues, soul mates, and, eventually, lovers. Einstein's parents strongly disapproved of Maleva and forbade him from marrying her. That only served to attract him to her even more. But much as he wished to marry Maleva, he simply couldn't.

Figure 3.2. Maleva Maric, Einstein's classmate, first wife, and mother of his three children

Einstein had no career and no income; he could not support a wife and family.

In 1901, Maleva became pregnant with Albert's child, a fact that he kept secret from his family and everyone else. Maleva returned to her Serbian home to deliver the child, a baby girl they named Lieseral. She left the baby with relatives and returned to Switzerland and Albert. Maleva and Albert succeeded in keeping Lieseral's existence completely secret throughout both of their lives. Only in 1986 did historians learn of her, three decades after her parents' deaths. No one really knows what happened to Lieseral; some believe she perished in a scarlet fever epidemic in 1903. But it is nearly certain that Einstein never saw his daughter.

Through his own actions, Einstein had firmly placed himself on a road to failure and obscurity. It seemed he might well fulfill the predictions of his many detractors and not merit even a footnote in the history of science.

PATENT CLERK

Figure 3.3. Einstein as a clerk in the Swiss patent office

Finally, in 1902, the father of Einstein's close friend Marcel Grossman pulled some strings in the Swiss government. Funds were allocated for a new entry-level position in the patent office in Bonn. The job description seemed to have been written to enhance Einstein's prospects. He applied for this job but was rated technically unqualified by the interviewer. However, the quest for civil service competency was trumped by political influence, and Einstein was hired as patent clerk third class—the bottom of the totem pole. Figure 3.3 shows an image of Einstein during his seven years as a patent clerk. Notice his attire—particularly bold and unconventional for the times.

In late 1901, Einstein submitted a Ph.D. thesis to Professor Alfred Kleiner of the University of Zurich, hoping a doctorate degree would improve his career prospects. Now that he had a job, he hoped a degree would secure his promotion to patent clerk second class and an increase to a more livable salary. However, his thesis was rejected.

THE WORST OF TIMES

In late 1902, at the nadir of Einstein's career, his father died. Hermann never lived to see his son achieve any of the success and celebrity that later came to the world's most famous scientist. Albert was profoundly despondent at his father's death, perhaps because he had never forged the bond with him that he so desired.

On his deathbed, Hermann finally relented and gave Albert his blessing to marry Maleva Maric. Despite his mother's continuing opposition, Albert and Maleva wed in January 1903. They were married for 15 years and had two more children, sons Hans Albert and Eduard. However, their marriage did not survive the fame and celebrity that Einstein ultimately achieved.

WAS PHYSICS FINISHED?

It is interesting to note the scientific climate of the times. At the end of the 19TH century, many felt that science and technology had reached the limits of what was possible. Very few saw opportunities for great discoveries ahead. One of the world's leading physicists of the day was England's Lord Kelvin, inventor of the Kelvin scale for measuring temperature. The Kelvin scale is often the most appropriate choice for scientific and engineering work because it defines zero degrees to be the coldest possible temperature, *absolute zero*, the temperature at which there is zero heat.

More on the Kelvin scale later. Lord Kelvin gave the keynote speech at the centennial celebration of Britain's Royal Society, the elite of British science, in which he said:

> "There is nothing new to be discovered in physics."
> — Lord Kelvin, 1900

The head of the U.S. Patent Office is said to have recommended that the government close the patent office and save the expense because he believed:

> "Everything that can be invented, has been invented"
> — Commissioner, U.S. Patent Office, 1899

Henri Poincaré, the leading French mathematician and physicist, said there were only three interesting problems left to be solved in physics. But these were seen more as loose ends rather than as opportunities for great discovery.

> Only Three Remaining Problems in Physics
>
> - Brownian motion
> - Photoelectric effect
> - Inability to find the luminiferous ether
>
> — Henri Poincaré, circa 1900

Isn't it amazing that the leading scientists of the day had no clue that the greatest discoveries in history were just around the corner?

Desperate to resurrect his failed academic career and facing a dearth of research topics, Einstein dedicated himself to solving all three problems identified by Poincaré, and to do so in spectacular style.

In 1905, Einstein succeeded, based solely on the sheer power of his genius, as we will discuss in the coming chapters.

NOTES

[1] There are many excellent biographies of Albert Einstein. Walter Isaacson's book *Einstein, His Life and Universe* is particularly comprehensive and enjoyable.

[2] As we will discuss in chapter 13, the resolution of this dilemma is two-fold. Firstly, with ever more powerful rockets, you can get ever closer to the speed of light—it's possible to reach 90% of the speed of light, or 99%, or 99.999%—but it's not possible to reach 100% of the speed of light, or more, under any circumstances. Thus you can never catch up to a light beam. Secondly, regardless of whether you are standing still, running along side a light beam at any possible speed, or even running away from it, light will always appear to be moving away from you at the same speed—671 million miles per hour—you can never see light travel at any other speed (in empty space). None of this is consistent with our everyday intuition, yet there is no longer any doubt that this is the way nature really is. One of the main reasons we all know Einstein's name is because he resolved this remarkable dilemma.

4

Einstein Settles Atomic Debate

While others attacked Boltzmann for his "unjustified" atomic assumption, Einstein focused on something he considered of great importance: the mathematical elegance of Boltzmann's theory, calling it "absolutely magnificent." Einstein had a deep-felt belief in the fundamental simplicity, harmony, and beauty of nature. He believed the laws of nature that scientists seek to discover are simple and elegant. Einstein was convinced that the elegance of Boltzmann's Statistical Mechanics reflected a profound reality—the existence of atoms—and he wanted to prove it.

Certainly the intellectual challenge excited Einstein, but there were also practical considerations. By 1904, Einstein was 25 years old and had failed to secure the academic job and professional status he so intensely desired. He needed a dramatic scientific success.

BROWNIAN MOTION

Let's examine how Einstein solved the 80-year-old mystery of *Brownian motion*. In 1827, the English botanist Robert Brown used a microscope to observe minute pollen grains suspended in a liquid. For no apparent reason, the pollen grains continuously and randomly darted this way and that. They moved one way and then would suddenly change direction as

Figure 4.1. Sample trail left by a pollen grain in Brownian motion, starting at the left arrow and ending at the right arrow

if they hit something invisible. Brown wasn't looking for exciting new physics. He wanted to study pollen and was annoyed that the grains kept moving. To him, Brownian motion was a nuisance. Some scientists suggested these bizarre motions might be due to collisions with atoms that were too small to be seen. But these suggestions were not quantitative and did not lead to a comprehensive theory to help interpret the observations. Additionally, the observations themselves were of little value because the motion of the pollen grains was very erratic, as shown in figure 4.1. Thus it was impossible to reach any definite conclusions on what caused Brownian motion, and the atomic debate continued.

WHAT IS SCIENCE? WHAT IS PROOF?

Humans are model-makers. We create mental models of the world around us that enable us to understand, anticipate, and prepare for nature's challenges and opportunities. Model-making is one of humanity's most distinguishing capabilities, and science is our collective effort to model nature. Models and theories are *effective* if they enable *predictions* that are of value and if the model is *validated*. Validation is essential but it is not the same as proof.

Scientific theories are never proven in the same sense as are mathematical theorems. In mathematics, once a theorem is proven by logical deduction, it is deemed TRUE and never needs to be proven again. For example, Euclidean geometry proves that the sum of the interior angles of any triangle equals 180 degrees. This is as true now as it was in Euclid's day 25 centuries ago. But mathematical logic cannot prove that this theorem is relevant to our universe. Euclid's theorem is about idealized triangles. Only experiment and observation can *validate* or *falsify* the hypothesis that the real world conforms to Euclidean geometry (in fact, the real world isn't always Euclidean). Extensive testing has verified that mathematics is generally a remarkably effective model of the real world, or else we wouldn't be able to build roads, skyscrapers, and airplanes.

However, models have limits and it's at the limits that interesting things occur. Scientists do not shrug off small discrepancies between observation and the predictions of our theories; they relish them. Even modest deviations can highlight deficiencies in current theories and lead us to better theories and a greater understanding of nature. And that, after all, is why people become scientists. The ancient Greeks observed small deviations of the real world from Euclidean geometry and realized that the Earth is round—one of science's greatest discoveries.

Thus scientific theories are never proven, once and for all. They are validated to a certain level of precision within a certain range of conditions and are subjected to the never-ending challenge of ever more precise testing in ever broader conditions. What makes a good theory? When do scientists accept a theory as an effective model of nature? The best theories are those which: (1) make the most predictions; (2) are the most precisely validated; and (3) are validated over the broadest conditions. It also helps if the predictions are remarkable. Predicting the Sun will rise in the east impresses no one. But when Einstein's prediction that the Sun bends starlight was validated, everyone was impressed.

Prediction and the ability to falsify hypotheses are the hallmarks of science that distinguish it from other endeavors. Richard Feynman, shown in figure 4.2, one of the most important physicists of the 20TH century and one of my professors at Caltech, said:

"The basis of science is its ability to predict."
— Richard Feynman, Nobel 1965

Figure 4.2. Richard Feynman (1918–1988) received the 1965 Nobel Prize in Physics for his contributions to Quantum Electrodynamics (QED). QED extends Maxwell's theory of electricity and magnetism into the micro-world of Quantum Mechanics, the subjects of later chapters.

To be a legitimate part of science, a theory or hypothesis must be testable—it must make predictions that can be confirmed or falsified. An idea that is impossible to disprove, or that does not predict anything we could possibly perceive, is not properly part of science.

For example, the statement "The universe has extra dimensions that are too small to ever be detected" cannot be falsified because it asserts that something exists which cannot be found. No experiment can disprove the existence of something that is undetectable. But the statement "The universe has extra dimensions that reduce gravity by 1% per billion miles" is falsifiable. This second statement makes a prediction that can be tested, even if our present instruments lack sufficient precision. If experiments ultimately find that gravity does not decrease as claimed, that statement would be falsified.

EINSTEIN FINDS SOLUTION

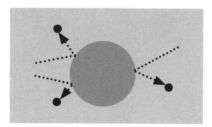

Figure 4.3. Pollen grain (gray) is hit by atoms (black) that are too small to be seen.

Einstein wasn't the only one to believe that Brownian motion is due to unseen atoms colliding with the much larger pollen grains, as illustrated in figure 4.3. But he was the first to conceive of an effective way to analyze these motions: study the forest rather than investigate each tree. Einstein realized it is better to measure the result of many collisions rather than try to measure and analyze each individual erratic motion. It is easier and more precise to measure how far an average grain travels in an extended time, such as one minute, during which many collisions occur.

Einstein derived predictions from his hypothesis. In 1905, he published a diffusion equation that predicts the average distance a pollen grain travels through a liquid in a given time [1]. The predicted distance depends on properties of the pollen and the liquid that are easy to measure. The distance also depends on the number of atoms in a given amount of liquid. With Einstein's equation, the number of atoms can be calculated from measurements of the average distance pollen grains travel and the properties of the pollen and liquid.

Einstein's hypothesis makes many falsifiable predictions. His diffusion equation predicts the average distance pollen grains travel for any values of four parameters: grain size, time, and the liquid's temperature and viscosity. For example, Einstein's equation predicts that raising the temperature 6% increases the average distance traveled by 3%. It also predicts that grains that are 4% smaller move 2% farther, with similar predictions for changes in travel time and viscosity. His equation predicts the change in distance for any amount of change in any of the four parameters. That's a lot of predicting. Einstein stuck his neck out quite far. If experiments had found a different distance for any combination of values of the four parameters, his hypothesis would have been falsified. It was not.

Einstein's predictions were confirmed in careful experiments by French physicist Jean Baptiste Perrin, who received the 1926 Nobel Prize in Physics for this work. Einstein had found the solution! His diffusion equation enabled the first definitive measurements of the mass and size of atoms. The 80-year mystery of Brownian motion and the 2500-year atomic debate were finally settled. Everything we see really is made of atoms.

But it turns out, what we call atoms are not the uncuttable *atomos* of Democritus.

NOTES

[1] Einstein's diffusion equation is much more complex than the typical equation in this book. The important part is: $d^2 = (some\ stuff) \times t/N$, where d is the distance the pollen grain travel during time t and N is the number of atoms in a specific volume of the liquid. We'll get to "some stuff" later. Increasing t allows the pollen to travel farther, thus it makes sense that t is in the numerator. The bigger N is, the more atoms exist in a specific volume, the smaller they must be, and the smaller their impact on pollen grains; thus N is in the denominator. Now d^2: if the pollen moved along a straight path with constant velocity v for time t the distance traveled would be $d = vt$. Instead, the grains are continually changing direction, often going sideways or even backward. Einstein's insight was to figure out how far they travel on average. This called a *random walk* or a *drunken sailor* problem. A drunken sailor falls down, gets up, takes a step in a random direction, and falls down again. Here, as Einstein found, it is the square of the distance that is proportional to time. It takes four times as long to travel twice as far when each step is taken in a random direction. The complete equation with all the "stuff" is: $d^2 = (yT/ru) \times t/N$, where y is a numerical constant, T is the liquid's temperature, r is the radius of the pollen grain, and u is the liquid's viscosity, which measures its resistance to motion. Molasses is more viscous than water, so u is larger in molasses.

5

Inside the Atom

Einstein proved that everything we see really is made of atoms, very small atoms. How small? A billion carbon atoms lined up in a row would be about one foot long. A 130-pound person contains 6 billion, billion, billion atoms [1]. As small as atoms are, we now know they are not the smallest components of matter. Atoms can indeed be cut into smaller pieces, and some of those pieces can be cut even further.

If we look inside an atom, as in figure 5.1, at first it seems to have only two parts: an inside and an outside.

The inside is a *nucleus*, which has a positive electric charge. The outside is a "cloud" of *electrons*, which has a negative electric charge. Opposite charges (positive and negative) attract one another, which holds atoms together.

The sketch is not nearly to scale. The electron cloud is really 100,000 times larger than the nucleus. If we imagine enlarging an atom until the nucleus becomes the size of a golf ball, the electron cloud would be over 2 miles wide. But even though it is so much smaller, the nucleus dominates the atom, containing almost all of an atom's mass—typically about 99.97%. Nuclei can also contain a million times more extractable energy than their surrounding electron clouds.

It takes energy to "cut" an atom into pieces—to pull electrons away from the nucleus—but now that's easy to do.

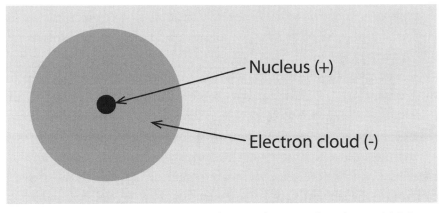

Figure 5.1. An atom has two principal parts: the central nucleus, which has a positive electric charge, and the surrounding diffuse cloud of electrons, which has a negative electric charge.

With even more energy we can pull nuclei apart. As shown in figure 5.2, nuclei are composed of *protons*, which have a positive electric charge, and *neutrons*, which have zero charge. Each nucleus must have at least one proton so its positive charge will attract electrons. The number of protons in an atom's nucleus is its most important characteristic: its *element number*. There are over 100 elements in the Periodic Table, each with different physical and chemical properties.

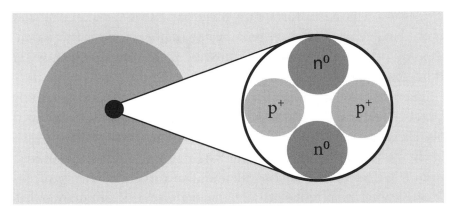

Figure 5.2. The nucleus of an atom is composed of one or more protons (p^+), which have a positive electric charge, and zero or more neutrons (n^0), which have zero charge.

Gold is element number 79 with 79 protons in its nucleus. Lead is element number 82 with 3 more protons than gold. What a difference 3 little protons make!

The number of neutrons is generally similar to the number of protons, but it can vary. Two atoms with the same number of protons and different numbers of neutrons are the same element and have the same chemical properties, but are different *isotopes*.

The most common atoms have relatively light nuclei and many of these have equal numbers of protons and neutrons. Carbon has 6 of each, nitrogen 7 of each, and oxygen 8 of each. In heavy nuclei, neutrons outnumber protons. In their most common isotopes, iron has 30 neutrons and 26 protons, while uranium has 146 neutrons and 92 protons.

Because atoms generally have the same number of protons and electrons, they are *neutral*—having zero net electric charge. Atoms with different numbers of protons and electrons have a non-zero electric charge—they are called *ions*.

Normal hydrogen has the smallest nucleus—one proton and zero neutrons. Deuterium is a rare, but important, isotope of hydrogen; its nucleus contains one proton and one neutron. While it's the smallest atom, hydrogen accounts for 92% of all atoms and 74% of all the atomic mass in the universe. Helium is element number 2; its normal isotope has two protons and two neutrons. Helium accounts for 7.5% of all atoms and 24% of all atomic mass. Together, hydrogen and helium make up almost everything we see in the cosmos: they account for 99.8% of all atoms and 98% of all atomic mass in the universe. Of the remaining mass, 1% is oxygen, 0.5% is carbon, and all other elements are much more rare.

On Earth, atomic abundances are quite different. Hydrogen is much less abundant than in the universe at large. About 10% of the mass of our oceans is hydrogen, but less than 0.2% of Earth's interior is hydrogen. Helium is almost totally absent on Earth; in fact, we first discovered helium in the Sun, and only later found it on Earth.

The atomic composition of the human body is very different from that of the universe, the stars, and the Earth. By mass, we are 65% oxygen, 18% carbon, 10% hydrogen, 3% nitrogen, 1.5% calcium, 1.2% phosphorus, and 1.3% other elements [2].

We are indeed made of fine and rare ingredients that were billions of years in the making, as we will explore later.

NOTES

[1] For simplicity, we will almost always round numbers off to just one digit. Since we will explore the very largest and the very smallest structures of nature, we must use some extremely large and some extremely small numbers. For these, we use the most common terminology:

1 million is 1,000,000	(6 zeros),
1 billion is 1,000,000,000	(9 zeros),
1 trillion is 1,000,000,000,000	(12 zeros).

We will also use these abbreviations:

2 M for 2 million

7 G for 7 billion.

A million, million is not one million *plus* one million, which is 2,000,000 (two million), but rather the much larger number: one million *times* one million, which is 1,000,000,000,000 (one trillion). Thus

6 billion, billion, billion is

6 followed by 27 zeros, which can be written as

6,000,000,000,000,000,000,000,000,000, or

6×10^{27}

Similarly, for small quantities, 3 billionths is 3 divided by one billion, which can be written in either of the following ways:

0.000,000,003 (the 3 is in the 9TH digit), or

3×10^{-9}

[2] Other important elements in the human body include: potassium, sulfur, chlorine, sodium, magnesium, iron, cobalt, copper, zinc, iodine, selenium, and fluorine.

6

Elementary Particles

We think we have finally completed a quest to understand the nature of matter that began over 2500 years ago.

Democritus was correct: matter is made of uncuttable pieces. But instead of *atomos*, we call these *elementary particles*. We are quite confident, but not absolutely sure, that elementary particles have no internal parts and are therefore the most basic building blocks of nature. Electrons are elementary particles. However, protons and neutrons are not elementary; each is made of three *quarks* that are elementary particles.

PARTICLES ARE EXACTLY IDENTICAL

One of the most important characteristics of elementary particles is that every particle of each type is *exactly* identical. For example, electrons are completely indistinguishable from one another. This does not mean that electrons are very similar, like two dimes or "identical twins." Nor does this mean that we lack instruments to adequately detect their differences. In fact, nature herself cannot discern any difference whatsoever between any two electrons in the entire universe. As we will discuss in the chapters on Quantum Mechanics, if there were even the smallest difference between electrons, atoms would collapse and life would not exist. All elementary particles of each type are exactly identical.

Even though protons and neutrons are not elementary particles, they are also exactly identical to all particles of their own types. This is because there is only one way to combine three quarks to make a proton and only one other way to make a neutron.

This characteristic of particles being exactly identical is remarkably different from anything we are familiar with in the macro-world.

FERMIONS: CONSTITUENTS OF MATTER

We divide particles into two groups: *fermions* and **bosons**. Bosons are the carriers of nature's forces and are discussed in chapter 8. Fermions are the building blocks of matter. They are named in honor of Italian-American physicist Enrico Fermi, my father's mentor, who is shown in figure 6.3. Fermi received the 1938 Nobel Prize in Physics for his discoveries regarding the physics of the nucleus, and was one of the most important physicists of the 20TH century.

Fermions have a quantum mechanical characteristic that makes them "antisocial." Fermions will not share their "turf" with any of their own kind. This antisocial trait gives matter rigidity. Material objects are difficult to compress because they are all made of fermions that resist being squeezed together. If not for this antisocial behavior, all Earth's atoms and all our bodies' atoms would collapse as gravity pulled everything together.

Elementary fermions are divided into two types: (1) *quarks*, which participate in the *strong nuclear force*; and (2) *leptons*, which do not. As we will discuss in chapter 8, the strong force is by far the most powerful force in nature.

As shown in figure 6.1, elementary fermions comprise a set of twelve particles: six quarks and six leptons. They are grouped into three *generations*, each containing two quarks and two leptons. We will discuss shortly the special use here of the word *generations*.

QUARKS

Particles with the strong force (Jedi?) are all made of quarks, the most elementary of the strongly interacting particles. The six quarks are named *up* (*u*), *down* (*d*), *charm* (*c*), *strange* (*s*), *top* (*t*), and *bottom* (*b*). The names are fanciful and have very little to do with their properties; none is more charming than the others, nor is any more strange.

Quarks combine to make hundreds of "larger" particles, the most important being the proton and the neutron. Two *up* quarks and a *down* quark make a proton; two *down* quarks and an *up* quark make a neutron.

CHARGED LEPTONS

Leptons are not affected by the strong force. They include the *electron* (*e*), *muon* (*μ*), and *tau* (*τ*), all with the same electric charge: –1. Leptons also include three types of *neutrinos* that have no electric charge and are discussed below.

Muons were first found among cosmic rays hitting Earth's surface. Once, muons were thought to be strongly interacting particles, and were even thought to be the basis of the strong force. But a famous experiment performed by Marcello Conversi, Ettore Pancini, and Oreste Piccioni (my father, shown in figure 6.2) proved that muons had no strong force interaction. The experiment was done in a high school basement in Rome during World War II, while bombs were falling on the Italian capital. This was before transistors, when vacuum tubes were the state of the electronic art. Dad said they preferred RCA tubes, but when these weren't available due to the war, they made their own vacuum tubes with hand-blown glass envelopes. Science historian Professor Robert Crease of Stony Brook University selected this experiment as one of the 20 most beautiful scientific experiments of all time. Dr. Daniela Monaldi, a science historian now at the Planck Institute in Berlin, said this experiment was the beginning of modern particle physics. The muon and tau are now understood to be essentially more massive versions of the electron.

	Generation Number			
	1	2	3	Charge
Quarks:	*u*	*c*	*t*	$+\frac{2}{3}$
	d	*s*	*b*	$-\frac{1}{3}$
Leptons:	*e*	*μ*	*τ*	−1
	v_e	v_μ	v_τ	0

Figure 6.1. The twelve elementary fermions are the most basic building blocks of matter. The six quarks combine to form all the particles that participate in the strong nuclear force. One combination of three quarks (uud) makes a proton while another combination (udd) makes a neutron. The six leptons do not participate in the strong force, and do not combine to form other particles.

NEUTRINOS

Neutrinos have no electric charge and come in three distinct types: v_e, v_μ, and v_τ—one type matched with each of the charged leptons. The Greek letter v is pronounced "nu." The properties of neutrinos have been well established, except for their masses. We know their masses are very small, no more than one billionth of a proton's mass. But because they are so light and so ghostly, we don't have precise mass measurements. This is in sharp contrast to the precise measurements we have for the masses of other particles—the mass of the electron, for example, is known with a precision better than one part in 10 million.

Neutrinos interact so weakly with other particles that they can easily pass through thousands of miles of steel without ever "hitting" anything. Detecting them and measuring their properties is a great challenge, requiring some of the largest and most-massive particle detectors. The Super-Kamiokande neutrino detector in Japan has a mass of 50,000 tons. (It's lots of fun, if you are not being paid per detected neutrino.)

Figure 6.2. Oreste Piccioni (1915–2002) received the 1998 Matteucci Medal. My father was renown for innovative experimental design, and for his discoveries regarding leptons, kaons, and antimatter. He pursued physics research in the U.S. for over 50 years and was a physics professor at UCSD.

FIRST-GENERATION FERMIONS RULE

As shown in figure 6.1, the twelve elementary fermions are grouped into three *generations*, each having two quarks and two leptons. Physicists don't use *generation* in the normal sense in this case (that would be too easy); none of the elementary particles are descendents of any others. Rather, particles of different generations are distinguished by their masses. While neutrino masses are uncertain, the masses of the quarks and charged leptons are all well known. The third-generation fermions are more massive than their second-generation counterparts that, in turn, are more massive than the first-generation. Because of their greater masses, second- and third-generation fermions lead very brief lives and are extraordinarily rare in nature. They are also much harder to produce

Figure 6.3. Enrico Fermi (1901–1954) received the 1938 Nobel Prize in Physics for his discoveries in nuclear physics. He was the last physicist who was both an outstanding experimentalist and an outstanding theorist. The largest particle accelerator laboratory in the U.S. is named in his honor, as is NASA's latest gamma-ray space telescope.

in our laboratories, and were discovered many decades after their first-generation counterparts (hence, the use of the word *generation*). For example, the electron was discovered in 1897; the muon, which is 207 times heavier, was discovered in 1937; and the tau, which is 3500 times heavier than the electron, was discovered in 1975. Because of their rarity, the heavier fermions are of little importance to the evolution of stars and the universe. Therefore, for the rest of this book, we will focus only on first-generation fermions—particles that have existed for 14 billion years.

Everything we see is made of first-generation charged fermions.

ARE THERE MORE PARTICLES?

We have good reason to believe, but no definitive proof, that there are no more elementary fermions beyond the twelve listed in figure 6.1. If a fourth fermion generation did exist, there should be a fourth type of neutrino. Recent experiments have searched for a fourth type of neutrino and found that if it does exist its mass must be billions of times greater than the masses of the known neutrinos. While that's not impossible, such a great mass difference between generations is considered highly improbable.

Some theorists have suggested the existence of additional types of particles, mostly bosons. Most notable is the Higgs boson that is thought to underlie the process that gives different particle types their different masses. But none of these hypothetical particles has yet been observed. If and when they are discovered and their properties are measured, they will be added to our list.

Thus the quest to find the most basic constituents of matter ends with the elementary fermions.

So much for matter.

But wait... there is also *antimatter.*

REFUGEES HELP U.S. FISSION WHILE FASCISTS FIZZLE

In the years leading up to World War II, many scientists, including 40% of all European theoretical physicists, fled fascist, anti-Semitic persecution. Among those who escaped were 14 physics Nobel Laureates, including Albert Einstein, Niels Bohr, and Enrico Fermi. It is widely believed the Nazis even offered a reward to anyone who killed Einstein, Europe's most celebrated Jew.

Receiving the Nobel Prize would be the crowning achievement of any physicist's career, but for Fermi it was even more important. His wife was Jewish, which put her life at risk. Fermi brought his wife and children to the Nobel Prize award ceremony in Sweden. After the ceremony, they escaped to the U.S. rather than return to Italy. His Nobel Prize may have saved her life.

In 1939, Einstein wrote a letter to President Franklin Roosevelt informing him that physicists still in Europe reported that the Nazis were trying to develop a bomb based on nuclear fission. Einstein told Roosevelt such a weapon could be devastating and recommended the U.S. start its own nuclear weapons program. Roosevelt launched the Manhattan Project, to which many refugee physicists made essential contributions. Fermi, who was the first to split the atom and the first to achieve a sustained nuclear chain reaction, played a major role in the Manhattan Project.

As we all know, the Manhattan Project succeeded and ended the war. Meanwhile, the German nuclear program failed, due in large part to a substantial miscalculation by its leader, Werner Heisenberg. He grossly overestimated how much uranium was needed to sustain an explosion.

How different the world might be if all of Europe's scientists had remained in Europe.

After the war, many German scientists were accepted back into the world scientific community—Wernher von Braun is an interesting example. However, Heisenberg was not welcomed back into the fold. For many, trying to develop nuclear weapons for the Nazis was too much to forgive.

7

Antimatter

In nature, for every yin there is a yang. For each type of particle of matter there is a corresponding type of particle of *antimatter*: an *antiparticle*. An antiparticle has the same mass as its particle partner, but all its other characteristics are the opposite of its partner's. Some particle types and their antiparticle partners are shown in figure 7.2.

For example, the *antielectron* has the same mass as an electron, but its electric charge is +1, whereas the electron's electric charge is –1.

Figure 7.1. Matter and antimatter are (almost) exact opposites.

Like the electron, all particles that have an electric charge have antiparticle partners that have the opposite electric charge, making the particle and the antiparticle distinctly different. Some electrically neutral particles are their own antiparticles. *Photons*, the particles of light, and "antiphotons" are one and the same thing, and are denoted by the Greek letter γ, pronounced "gamma." Some other neutral bosons, such as the Z^0 and the *graviton*, are also their own antiparticles.

Particle	Mass	Charge	Symbol
Proton	1836	+1	p^+
Antiproton	1836	−1	p^-
Neutron	1839	0	n^0
Antineutron	1839	0	$\overline{n}^0$
Electron	1	−1	e^-
Antielectron	1	+1	e^+
Neutrino	$<10^{-6}$	0	ν, 3 types
Antineutrino	$<10^{-6}$	0	$\overline{\nu}$, 3 types
Photon	0	0	γ

Figure 7.2. These are the most common particles of matter and their anti-matter partners, with relative masses and electric charges. Partners have the same masses, but all their other properties are opposite.

Neutrons and *antineutrons* both have zero charge, but they are different particles. Neutrons are made of quarks and antineutrons are made of *antiquarks*. The electric charges of antiquarks are opposite to those of quarks. Since quarks and antiquarks are different, so are neutrons and antineutrons.

The key point is that all the characteristics of a particle and its antiparticle exactly cancel one another. If a particle and its antiparticle combine, they *annihilate*—they totally destroy one another and leave behind only energy.

This process is illustrated in figure 7.3, where an electron and an antielectron collide and annihilate. Their mass is converted into energy that in this example is carried away by two photons. All the characteristics of a particle-antiparticle combination sum to zero—matching the value of energy. When such pairs annihilate, the original particles cease to exist.

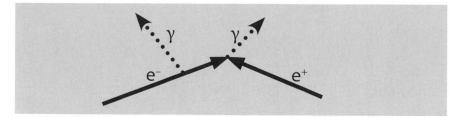

Figure 7.3. Feynman diagram illustrates pair annihilation. An electron e⁻ and an antielectron e⁺ (solid lines) enter at the bottom and annihilate when they meet in the center. In this case, their energy is carried off by two photons γ (dotted lines) that exit at the top. Feynman diagrams are discussed further in chapter 8.

After annihilation, no trace remains of the particle or the antiparticle—no residual charge, no residual matter, no residual anything.

The reverse process is also possible. Any particle-antiparticle pair can be created from pure energy. This is the bread-and-butter of experimental particle physics. By smashing two particles together often enough and hard enough, we eventually create every particle nature allows, with its antiparticle. We don't even have to know what we are looking for—particles just appear. If the available energy in a collision is E, we can create any particle-antiparticle pair whose combined mass is less than E/c^2 (this is due to Einstein's equation $E=mc^2$, which we will discuss in chapter 9). It's no wonder physicists always want higher-energy particle accelerators—all the better to cook up new particles.

Particle-antiparticle pair creation is illustrated in figure 7.4. Here, an antineutron and a neutron are created from the collision energy of two protons. The antineutron was discovered by a team led by Oreste Piccioni (shown in figure 6.2); they used the reaction shown in figure 7.4 to produce them.

In almost all cases, any interaction that can occur among a group of particles can also occur among the corresponding group of antiparticles. Nature does not favor either particles or antiparticles; this is called **CP-symmetry**. However, there is a very important exception. A number of experiments, including my Ph.D. thesis, demonstrated that in certain special situations CP-symmetry is very slightly *broken*. (Note the slight

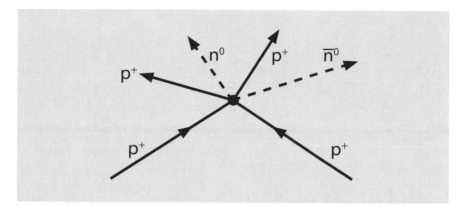

Figure 7.4. Pair creation is illustrated above. Two protons (p^+) enter at the bottom and collide in the center. Exiting the collision are the original protons plus two new particles: a neutron (n^0) and an antineutron ($\overline{n}^0$) that are produced from a portion of the kinetic energy of the original protons.

asymmetry in figure 7.1.) Breaking CP-symmetry means certain processes can create more particles than antiparticles. This played a critical role in the evolution of our universe by allowing slightly more matter than antimatter to develop during the first one second of existence. The amount of excess matter was very slight indeed: for every one billion antielectrons there were one billion *and one* electrons.

After the universe was one second old, it was too cold [1] for new particle-antiparticle pairs to be copiously created in collisions, such as the one in figure 7.4. Antimatter annihilated with matter, creating a tremendous number of photons and leaving behind the slight excess of matter that had developed earlier. The antimatter ran out before the matter did and today there is almost no antimatter left in the universe. Everything we see is made of that slight excess of matter.

NOTES

[1] "Too cold" means particle kinetic energies were too low. At one second, the actual temperature was 10 billion degrees.

Four Forces of Nature

We know of these four forces of nature:

Strong	Holds quarks together in protons and neutrons. Holds protons and neutrons together in nuclei.
Electromagnetic	Holds together atoms, molecules, and larger bodies, including humans.
Weak	Causes radioactive decay.
Gravity	Holds together massive bodies.

WHAT IS A FORCE?

Sir Isaac Newton established the concept of forces as agents that change velocity. His first law states a principle that was discovered by Galileo: without a force, velocities do not change; objects at rest remain at rest, and objects in motion remain in motion. His second law states that when a force is applied to an object its velocity changes—it *accelerates*:

Force = (*mass*) multiplied by (*acceleration*)

or more simply, $F = ma$.

Acceleration is how rapidly velocity changes. For example, top dragsters change their velocities from zero to 330 miles per hour (mph) in 4.5 seconds. That's an acceleration of 74 mph per second, over 3 times the acceleration of gravity on Earth's surface.

How do forces work? It's not hard to understand how a bat applies a force to a ball—through direct contact. At a microscopic level, "contact" means atoms in the bat are squeezed against those in the ball. The electrons in these atoms repel one another, thus "forcing" the ball and bat apart. But Newton was unable to explain, as was everyone else before Einstein, how the Sun could exert a force on Earth across 93 million miles of empty space. This mysterious effect was called *action-at-a-distance*. No one, including Newton, was comfortable with an unexplained, non-contact force, but because his equations gave the right answers for the motions of apples and planets, they were accepted and held sway for over two centuries.

The modern view of forces is: (1) the force between any two objects is the sum of the forces between all the individual particles of one object and all the individual particles of the other object, and (2) the force between any pair of particles arises because they exchange another type of particle called a *boson*, as illustrated in figure 8.1.

This modern view of forces comes from a branch of physics called Quantum Field Theory and incorporates the concept of *virtual particles* that we will discuss in chapter 17 on Quantum Mechanics.

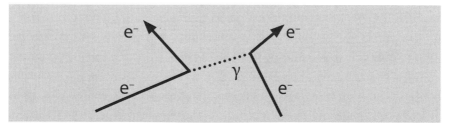

Figure 8.1. Feynman diagram showing two particles (in this case, electrons e⁻) exchanging a boson (in this case, a photon γ). This process changes the electrons' velocities and results in an "electromagnetic force."

Force	Associated Bosons
Strong	gluons
Electromagnetic	photon, γ
Weak	W^+, W^-, Z^0
Gravity	graviton (not yet observed)

Figure 8.2. The four forces and their associated bosons

BOSONS: CARRIERS OF FORCES

Bosons are named in honor of Indian physicist Satyendra Nath Bose, who, in 1924, described a key quantum mechanical aspect of the behavior of these force-exchange particles. Bose's ideas were quite novel, and as he was not known and not European, scientific journals declined to publish his work. Refusing to be denied, Bose sent his paper to Einstein requesting his endorsement. Einstein saw great value in Bose's ideas. He translated Bose's paper into German, wrote his own supplementary paper expanding on Bose's ideas, and ensured both papers were published together. These principles are now referred to as *Bose-Einstein statistics*.

The strong, electromagnetic, and weak forces each have their own bosons, listed in figure 8.2. Quantum Field Theory deals with these three forces very effectively, but has not yet been successfully extended to gravity. Field theory presumes an exchange boson for gravity, named the *graviton,* but that has not yet been detected, and calculations based on graviton exchange do not yet yield sensible results. Einstein's theory of gravity, General Relativity, is very effective for large, *macro-world* objects, but it is not a quantum theory as is required in the *micro-world*. Closing this theory gap is one of the highest priorities of modern physics. Only with an effective theory of gravity that works both in the macro-world and in the micro-world will we better understand the center of *black holes* and the moment of the *Big Bang*.

The schematic shown in figure 8.1 provides insight into the physical process and is a starting point for quantitative analysis. It is called a

Force	Strength
Strong	10,000
Electromagnetic	100
Weak	1
Gravity	7×10^{-34}
= 0.000,000,000,000,000,000,000,000,000,000,000,7	

Figure 8.3. While force strengths are not directly comparable, this chart provides a useful prospective. The weak force is assigned a strength of 1.

Feynman diagram in honor of its inventor, Nobel Laureate Richard Feynman. (It was a privilege for me to learn "The Diagrams" from Feynman himself while I was a student at Caltech.) In this figure, two electrons enter from below and interact by means of the electromagnetic force. The left electron emits a photon that the right electron absorbs; the photon changes the velocities of both electrons, resulting in a "force." The right electron can also emit a photon that the left electron can absorb. In fact, there are innumerable diagrams of ever greater complexity for the interaction of two electrons. Feynman's rules make it easier to systematically list the possible diagrams, identify the important ones, and compute their effect.

The four forces have radically different strengths, although a precise comparison isn't possible. Because the forces vary differently with distance and particle type, comparing their strengths is a little like comparing apples and oranges. But a general feeling for their relative strengths is shown in figure 8.3. Here I have arbitrarily assigned a strength of 1 to the weak force. In those terms, the strength of the strong force is 10,000 and the strength of the electromagnetic force is 100. Gravity is immensely weaker; its relative strength on our scale is 7×10^{-34}. If you are not familiar with this notation, this is a number vastly smaller than 1. When written as a decimal fraction, as it is in figure 8.3, it has 33 zeroes followed by a 7 in the 34TH digit. Gravity's strength is less than a trillionth of a trillionth of a trillionth of that of the strong force.

Let's now discuss each force more fully.

STRONG FORCE

As figure 8.3 shows, the strong force earns its name because it is by far the strongest of nature's forces. It holds together quarks within protons and neutrons, and it holds together protons and neutrons within atomic nuclei. Because the strong force holds these particles together so tightly, its exchange bosons are called *gluons*. The strong force underlies *nuclear fusion*, which powers the stars and creates the energy that sustains life. Nuclear fusion also produced the atoms from which Earth, this book, its author, and you the reader are made.

In our everyday lives, we don't notice the strong force because its range is extremely short, only about the size of a proton, not even as large as many atomic nuclei. Unless two nuclear particles almost touch, they don't feel the attraction of the strong force. However, if they do get close enough, the strong force dominates all others; it holds protons together in nuclei even though they all have a positive electric charge and positive charges repel one another.

ELECTROMAGNETIC FORCE

The electromagnetic force is the next strongest force. It was once thought that electricity and magnetism were two different forces. However, in the 19TH century, English physicist Michael Faraday and Scottish physicist James Clerk Maxwell developed the theory of electromagnetism, providing a unified concept of electricity, magnetism, and light.

Faraday was born in humble circumstances at a time when science was primarily an activity for the upper class. Largely self-taught, he slowly but steadily worked his way up from technical assistant to professor of the Royal Institution. Faraday compensated for his poor math skills with creativity, meticulous experimentation, and great physical intuition. He discovered magnetic induction and built the first electric

Figure 8.4. Michael Faraday (1791–1867) and James Clerk Maxwell (1831–1879), right, were the principal discoverers of electromagnetic theory.

motor. Faraday visualized electricity and magnetism in terms of field lines. This concept ultimately led to the development of the theory of fields, the foundation of most modern physical theories. He discovered that magnetic fields affect light beams, and proposed that light was an electromagnetic phenomenon.

Maxwell was born into a wealthy family, some 40 years after Faraday. Well-educated and particularly strong in mathematics, Maxwell created a comprehensive theory from all the bits and pieces that had already been discovered about electricity and magnetism.

Ultimately, Maxwell showed that waves of electromagnetic energy travel through space at a fixed speed that happens to match the measured speed of light. Near the end of Faraday's life, Maxwell visited him to deliver the news that light was indeed electromagnetic, as Faraday had earlier surmised.

Images of Faraday and Maxwell are shown in figure 8.4. Their discoveries remain valid to this day, and we now recognize electricity and magnetism as being two sides of one coin.

The electric force can be attractive or repulsive. Two electric charges with the same polarity (both positive or both negative) repel one another.

But protons and electrons attract one another because they have opposite electric charges (+1 and –1 respectively). This is the force that holds atoms together, binding electrons to nuclei. It also holds atoms together to form *molecules,* and it holds together complex structures such as the human body. Electromagnetism underlies all of chemistry and biology, and most of engineering. This force has an infinite range because its exchange boson—the *photon*—has zero mass. The strength of the force between two charges decreases rapidly with increasing distance (dropping 4-fold when the distance doubles), but it never drops to zero. In principle, electrons at opposite ends of the universe repel one another with an infinitesimal force. In practice, this is not important, in part because large objects have a strong tendency to be electrically neutral— to have the same number of positive and negative electric charges. If a large object, such as the Sun, had an unbalanced charge, it would either emit its excess charges or attract opposite charges from its surroundings until it neutralized itself.

WEAK FORCE

The weak force is responsible for radioactive decay that provides much of the heat in Earth's interior and keeps its iron core molten. Thanks to a molten iron core, Earth has a robust magnetic field that shields our atmosphere from erosion by the solar wind—charged particles streaming out of the Sun. By comparison, with less internal heat, Mars has a solid core and little or no magnetic field. Its atmosphere was blown away by the solar wind long ago. The weak force also allows, in certain conditions, protons to convert into neutrons and neutrons to convert into protons. This enables the production of all elements other than hydrogen and is therefore essential to life. Since the three exchange bosons of the weak force are all very massive, it has a very short range—only about 1% of the size of a proton. It's exceedingly unlikely that two particles ever get "close enough" to interact via the weak force. Reactions that proceed only through the weak force, such as radioactive decays and neutrino interactions, occur at extremely low rates.

GRAVITY

Last is gravity, the weakest yet most important force of all. Its relative strength on our scale is 7×10^{-34}. Thus the question arises, "How can such a feeble force be so important?" It is because gravity has an infinite range and is attractive for all forms of matter. The very short ranges of the strong and weak forces limit particle interactions to only a few close neighbors. Electromagnetism has an infinite range, but cancels at large distances because large objects are uncharged. Only gravity has an infinite range and is attractive for all types of matter, hence gravity keeps accumulating. The Sun contains 10^{57} particles (1 followed by 57 zeroes). Every one of those 10^{57} particles exerts its gravity on every one of the 10^{51} particles in the Earth. Even though the gravitational force between two particles is incredibly small, all these tiny forces add up—all 10^{108} of them—making the Sun's force on Earth huge.

Thus despite its intrinsic weakness, gravity dominates all other forces on the scale of planets, stars, galaxies, and the universe. It holds us on the ground as Earth turns, makes planets round, shapes galaxies, and holds together galaxy clusters. Gravity shapes everything in the cosmos and determines its fate. That makes it the most important force. Look at the impressive effect of gravity in Color Plate 4 and in figures 8.5 and 8.6.

Color Plate 4 is an image of **M51**, the majestic Whirlpool Galaxy, and its smaller companion galaxy **NGC** 5195. At 65,000 *light-years* across and with 160 billion stars, the Whirlpool is about half the size of our own galaxy, the Milky Way. A *light-year* is a convenient measure of astronomical distances. It's the distance light travels in one year, which is about 6 trillion miles. The speed of light is a very important topic that we will discuss in coming chapters. For now, what you need to know is that the speed of light (in empty space) is always the same and that a light-year is a unit of distance, not a unit of time. The Whirlpool is 23 million light-years distant and lies half a light-year in front of **NGC** 5195. These galaxies are gravitationally bound to one another and it appears the smaller galaxy has passed through the larger one within the last 100 million years.

Let's take a moment to discuss how celestial objects are named. Often their names derive from their sequence in astronomers' catalogs.

Figure 8.5. Left: Image by Jean-Charles Cuillandre, CFHT, of star cluster M22 that contains over 100,000 stars and is one of over 10,000 major star clusters within our galaxy. M22 is 10,000 light-years away and is 100 light-years across. Right: Image by Yuri Beletsky, ESO, of the Large Magellanic Cloud (LMC) in the southern sky that was first observed by Europeans during Magellan's circumnavigation. LMC is a satellite galaxy of our Milky Way; it is 168,000 light-years away and has a total mass of 10 billion Suns.

"NGC" refers to the New General Catalog published in 1888 by Danish-Irish astronomer John Dreyer. Initially it contained 8000 cosmic objects and was based on observations by German-British astronomer Sir Frederick William Herschel. Objects M1 through M110 were cataloged by Charles Messier, an 18TH century French astronomer, whose special interest was comet hunting at a time when that was *très chic*. To save time, Messier cataloged every fuzzy celestial object that turned out *not* to be a comet. Messier's list of 110 non-comets comprises a treasure trove of the brightest star clusters, nebulae, and galaxies. These continue to intrigue astronomers, both professional and amateur, even those with modest telescopes. Each spring, amateur astronomers delight in the "Messier Marathon", an attempt to view all 110 Messier Objects in one night.

In this chapter, I have included gravity as a force, as that is the traditional and the simplest way to begin. Later, we'll discuss Einstein's vision that gravity is better described as the *curvature* of space and time.

Figure 8.6. The immense galaxy cluster Abell 1689 is 2.2 billion light-years away, spans 2 million light-years, and contains trillions of stars. Each fuzzy spot is a giant elliptical galaxy. Image by Benitz, ESA, NASA Hubble.

Gravity
pulls galaxies together across a billion, billion miles
and pulls apples off trees here on Earth.

9

Energy, Mass, and $E=mc^2$

In 1905, Einstein revolutionized our understanding of mass and energy by publishing the most famous of all equations: $E=mc^2$. Before Einstein, physicists thought they understood *mass* and *energy*, and they believed mass and energy were two distinct, unrelated phenomena. Einstein saw much deeper than anyone had seen before, and showed physicists their beliefs were wrong. He saw an underlying unity between mass and energy, and his vision has forever changed our lives and our world.

In chapters 9 and 10, we discuss how Einstein achieved this unified concept of mass and energy, and learn how we can use that new understanding to address a very real and practical problem: How can we continue to obtain the energy we voraciously consume without destroying our environment?

First, what are mass and energy?

Mass is a measure of how much matter something contains. Consider the three components of atoms: protons, neutrons, and electrons. Protons and neutrons have about the same mass, and both are about 2000 times more massive than electrons. We can, therefore, ignore electrons for now and say that an object's mass is simply a count of how many protons and neutrons it contains. Mass is different from weight. A hammer weighs less on the Moon, where gravity is less, and it weighs nothing in the International Space Station, where there is zero gravity. But the hammer always has the same mass regardless of its location because it always contains the same number of protons and neutrons.

What is energy? Physicists have precise technical definitions of the various types of energy. But perhaps it is more informative to say energy is how we measure existence—energy is the currency of existence [1]. Everything that has physical existence has energy, and we can measure how much of something there is by measuring how much energy it has. Finally, and very importantly, energy is *conserved*, which means the total amount of energy never changes. Energy cannot be created nor destroyed; it can only be converted from one form to another.

Energy comes in many forms: work, heat, kinetic energy, potential energy, and many others. Physicists have long known that energy can be converted from one form to another (with some limitations). For example, consider the kinetic energy and potential energy of a pendulum as illustrated in figure 9.1. When the pendulum comes to the top of its swing, it stops for an instant before reversing direction and falling back down.

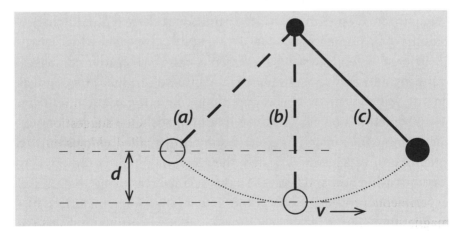

Figure 9.1. As a pendulum moves from the top of its swing (a) to the bottom (b), dropping a distance d, its potential energy is converted to kinetic energy and its velocity v reaches its maximum. As the pendulum continues swinging up to the top again (c), its kinetic energy is converted to potential energy. Because energy is conserved, the total of its potential and kinetic energies never changes—as one gets larger, the other must get smaller.

When it stops, its ***kinetic energy*** is zero because this is the energy of motion, but its ***potential energy*** is at its maximum. Gravitational potential energy is the energy an object has when it is elevated in a gravitational field; it takes energy to lift an object up against gravity. (That becomes more obvious the older we get.) Thus the pendulum's potential energy at the top of its swing is how much energy is needed to lift the pendulum up a distance ***d***. When the pendulum reaches the bottom of its swing, its potential energy reaches its minimum value, while its velocity—and hence its kinetic energy—reaches its maximum value. At the top of its swing, the pendulum's energy is entirely potential, while at the bottom, potential energy has been converted to kinetic energy. As it continues swinging, kinetic energy is converted back to potential energy. Because energy is conserved, the total of the pendulum's potential and kinetic energies never changes. Thus the change in potential energy is always opposite to the change in kinetic energy; when one goes up the other must come down. When we calculate a pendulum's velocity at any height from the equations for kinetic and potential energy [2], we find that the velocity does not depend on the pendulum's mass. The time it takes to swing through one full cycle is the same for both heavy and light pendulums of the same length.

Before Einstein, physicists thought mass and energy were completely different, and they thought mass was conserved in the same sense as energy was. But no one had figured out where the heat comes from when something burns, like a log in a fireplace. One suggestion was that combustible materials contain something called ***phlogiston***, the essence of fire. As a log is heated, they imagined phlogiston pouring forth releasing heat and flames. All this seemed reasonable, but careful experiments proved the phlogiston idea wrong. Some materials, like magnesium, gain weight when burned, which is hard to jibe with the notion of phlogiston flowing out.

In 1905, Einstein shocked the scientific world by asserting that mass was another form of energy. Mass seems nothing like any previously known form of energy, such as heat or kinetic energy. Those other forms are intangible—you cannot touch kinetic energy. Mass is very different; it is the epitome of being tangible. Heat can be easily created (just light a

Figure 9.2. Was it as easy as *a*, *b*, *c*? An apocryphal image of how the most famous equation of science might have been discovered.

candle), but no one had ever seen the creation of mass. Yet Einstein said these were all forms of the same thing—energy. And, he said, mass was not conserved but could be converted into other forms of energy and other forms of energy could be converted into mass. What is conserved is the total amount of energy in all its forms, including mass.

The meaning of Einstein's famous equation is similar to what all of us know about money. We know there are many forms of money: dollars, pesos, euros, yen, etc. We know we can convert money from one form to another, as shown in figure 9.3. Tourists to Japan may wish to convert dollars into yen. They can easily find out how many yen their dollars are worth: just multiply the number of dollars times an exchange rate to get the number of yen. (The first time I went to Japan, a U.S. dollar was worth 380 yen. How times have changed!)

Einstein said something very similar about mass and energy. Mass can be converted into another form of energy, such as heat. The amount of heat energy E equals the amount of mass m multiplied by an exchange rate c^2. Here c represents the velocity of light in empty space. It's a very big number—671 million mph—the fastest speed that anything can travel through space, and c^2 is c times c, which is a very, very big number.

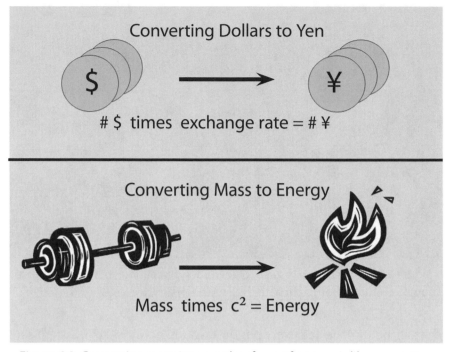

Figure 9.3. Converting mass into another form of energy is like converting dollars into yen. The exchange rate for mass conversion is c^2.

Thus even a small amount of mass converts into a huge amount of energy in another form.

As an example, consider the single U.S. penny shown in figure 9.4. If the mass-energy in a penny were converted into useful work energy (we will discuss how this might be possible later), it would provide all the energy requirements of a million people for a day! That is as much energy as is released from burning two million gallons of gasoline. One might say that the intrinsic value of gasoline is two million gallons for one penny.

We now know that almost all the energy we consume ultimately comes from converting mass into another form of energy, such as heat. Even solar cells and wind turbines, which are driven by energy from the Sun, are ultimately powered by mass conversion, because that is what powers the Sun and all the other stars.

Our Sun's heat and light come from the conversion of mass through a process called *nuclear fusion*, which we will discuss in the next chapter. Each second, the Sun "burns" over half a billion tons of hydrogen and converts 4 million tons of mass into heat and light.

However, there's no need to worry about our Sun burning out any time soon. In a billion years, the Sun consumes only 1% of its hydrogen gas—it's that immense.

Figure 9.4. The mass energy in a penny is equal to the chemical energy in 2 million gallons of gasoline, according to Einstein's equation *E=mc²*.

> Every second, our Sun "burns" 574 million tons of hydrogen, creating 570 million tons of helium, and converting 4 million tons of mass into energy.

NOTES

[1] Energy and existence are sometimes *potential* rather than actual. A rock at the top of a hill has potential energy. If it falls to the valley below, that potential energy is actualized.

[2] Near Earth's surface, a pendulum's potential energy = mgd, where m is the pendulum's mass, d is its height above its lowest point, and g is the acceleration of gravity on Earth's surface. Its kinetic energy is $\frac{1}{2}mv^2$. Equating the change in potential energy to the change in kinetic energy yields $\frac{1}{2}mv^2=mgd$, or $v^2=2gd$.

10

Smart Energy

One day, mankind will look back in amazement at today's energy technologies. Why were we so primitive? So wasteful? So destructive? Why was energy so expensive?

In the last chapter, we discussed Einstein's most famous equation $E=mc^2$ and its implications for generating energy, namely that almost all useful energy ultimately comes from the conversion of mass into other forms of energy, such as heat. With that perspective, let's now consider how to best supply our large and growing energy needs without destroying our environment. We will round off most numbers for convenience.

In 2004, total world energy consumption was 140 trillion kilowatt-hours (kWh), where energy consumed in all forms has been converted into electrical units. Average per capita energy consumption varied widely by country, with Canada topping the list at 480% of the world average, followed closely by the U.S. at 460%. Some other notable figures are: Australia at 330%, the U.K. at 220%, Italy at 180%, and China at 70%. Most industrialized nations ranged from 200% to 300% of the per capita world average. The sources of the world's energy were: 37% from oil, 25% from coal, 23% from natural gas, 6% from nuclear fission, 4% from biomass, 3% hydroelectric, ½% solar, and ⅓% from wind.

Providing the world's energy needs in 2004 required converting 6 tons of mass into other forms of energy. The U.S. share was about 1 ton.

As that's a nice round number, let's work with that and discuss the challenge of how to best convert 1 ton of mass into usable energy. For brevity, we'll say we need to generate "1 ton of energy." What are our options?

Options to Supply 1 Ton of Energy		
Process	Tons of Fuel Needed	Clean?
Burn Coal	5,000,000,000	No
Burn Gasoline	2,000,000,000	No

FUEL-BASED ENERGY

Let's first look at fuel-based energy sources, beginning with coal. We have a large supply of coal, it is cheap, and it produces a lot of heat. That's why most of our electricity comes from burning coal. To get 1 ton of energy, we would need to burn 5 billion tons of coal. Careful weighing of everything before and after burning all that coal would reveal that 1 ton of mass was lost during combustion. It is the conversion of that 1 ton of mass that produces all the heat and other energy released from burning that coal. Clearly, this is a very inefficient process; it takes 5 billion tons of coal to get just 1 ton of energy. It is also a very dirty process. In addition to the human and environmental costs of mining 5 billion tons of coal, the combustion creates billions of tons of polluting waste. Can't we do better?

We could burn gasoline; it has more pop per pound than coal. It turns out that we need to burn 2 billion tons of gasoline to get 1 ton of energy. Better, but this still produces billions of tons of pollution, and petroleum is a limited and expensive resource.

Why can't we burn coal or gas much more efficiently? Clever engineers have improved electronic circuit performance a billion-fold, over several decades. Why can't we at least triple the efficiency of coal or gas combustion? If we were three times more efficient, we would eliminate ⅔ of our costs, pollution, and consumption of precious resources.

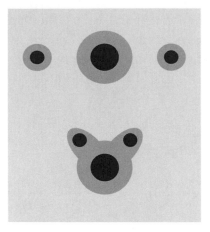

Figure 10.1. Two hydrogen atoms and an oxygen atom (above) react to form water (below). Chemical reactions such as this are essential to life, but their energy output is limited because only the electrons change their state. Nuclei are unaffected by chemical reactions.

Certainly we can do better; there are inefficiencies that can and should be eliminated. However, even with clever engineering and serious conservation, we will never overcome a fundamental physical limitation of all conventional fuels. Burning oil, coal, gas, biomass, etc. are chemical reactions that will always be grossly inefficient. This is because, at a fundamental level, in any chemical reaction, atoms change the way they share their electrons, but their nuclei remain unchanged. Figure 10.1 shows an example: two hydrogen atoms react with a larger oxygen atom to produce a molecule of water (H_2O). In this reaction, electrons drop to lower energy levels and release their excess energy, which we perceive as heat. This is one of the most potent chemical reactions, potent enough to blast rockets into outer space. But its power is limited because electrons have such a small share of an atom's energy. As discussed in chapter 5, nuclei can have 99.97% of an atom's mass and contain millions of times more extractable energy than the surrounding electrons.

To extract much more energy, to become much more efficient, we must go where the energy is—in the nucleus. There are two ways to change a nucleus and extract energy: (1) make it smaller, or (2) make it larger. (And you thought nuclear physics was complicated.)

In *nuclear fission*, the nucleus becomes smaller. As figure 10.2 illustrates, a very large nucleus, such as uranium, can split, or we can cause it to split, into two medium-sized nuclei. This is the process used in all nuclear power plants in operation today. Fission can release 100,000 times more energy than a typical chemical reaction. Unfortunately, the end products of nuclear fission are unstable and radioactive. To get 1 ton of energy from nuclear fission, we could "burn" 50,000 tons of uranium-235. The good news is fission produces no greenhouse gases and is 100,000 times more efficient than burning coal, thus 100,000 times less waste is produced. The bad news is no one wants 50,000 tons of radioactive material buried in their backyard.

What about making the nucleus bigger? That is *nuclear fusion*, illustrated in figure 10.3. Two or more small nuclei merge to produce a larger nucleus. Nuclear fusion can release millions of times more energy than a typical chemical reaction, and its end products are generally benign. In fact, much better than benign, the end products of nuclear fusion include carbon, oxygen, nitrogen, and iron, all elements essential to life. Every atom of these vital elements in the entire universe was produced by nuclear fusion in the centers of massive stars. We would not exist without nuclear fusion.

Can we generate energy with nuclear fusion? Not yet, but we can and should learn how. One day we will be able to "burn" water! Don't tell your favorite fire department, but water is 500 times more flammable than gasoline (?). What I mean is water contains an important but rare substance called *heavy water*.

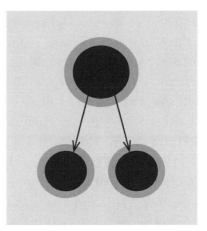

Figure 10.2. In nuclear fission, a very large nucleus divides into two smaller nuclei, releasing 100,000 times as much energy as a typical chemical reaction.

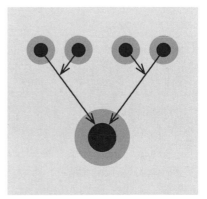

Figure 10.3. Nuclear fusion of four hydrogen nuclei to form a helium nucleus releases millions of times as much energy as a typical chemical reaction. Nuclear fusion produces carbon, oxygen, nitrogen, iron, and other elements vital to life.

In heavy water, a normal hydrogen atom is replaced by a different isotope of hydrogen—deuterium. Deuterium is the easiest substance to fuse and may therefore lead us to the first nuclear fusion reaction we master for energy generation. Deuterium is twice as heavy as normal hydrogen, thus heavy water sinks. Once we master fusion, we will be able to pump 4 million tons of water from the deep ocean, separate the heavy water, extract its deuterium, and fuse that deuterium to get 1 ton of energy.

What about pollution? At the end of this process, we will have over 3,999,300 tons of normal (non-heavy) water that can be returned to the ocean. We will also have 500 tons of breathable oxygen and over 100 tons of perfectly clean helium gas. We could use the helium to fill party balloons and blimps, or release it into the air where it would float to the top of the atmosphere and escape into outer space. Since 24% of the cosmos is helium, that wouldn't even pollute space. If we performed the process perfectly, we would be left with nothing else—no greenhouse gases and no pollutants—a perfectly clean process. We will be able to get abundant, clean energy from under the seas rather than having to buy it from overseas. Imagine how that will transform geopolitics!

Process	Options to Supply 1 Ton of Energy Tons of Fuel Needed	Clean?
Burn Coal	5,000,000,000	No
Burn Gasoline	2,000,000,000	No
"Burn" Seawater	4,000,000	yes?
Uranium Fission	50,000	No
Hydrogen Fusion	133	yes?
Mass → Black Hole	1	Yes

Can we do even better? As our technology advances, we will move beyond deuterium fusion to hydrogen fusion, the process that powers the stars. Only 133 tons of hydrogen is needed to generate 1 ton of energy, and the only residue of hydrogen fusion will be 132 tons of perfectly clean helium. This is 40 million times more efficient than burning coal!

We will have a virtually unlimited supply of clean energy from hydrogen in the ocean. Nuclear fusion is not some pie-in-the-sky theoretical dream; it is a well understood, widely occurring, natural phenomenon without which we wouldn't exist. Just as civilization took a great leap forward when we long ago learned to master fire, we will take another forward leap when we learn to master nuclear fusion.

But there is an even more efficient, long range alternative, due to British physicist Sir Roger Penrose, shown in figure 10.4. An advanced civilization could be positioned in orbit around a suitable *black hole*, or perhaps its scientists could learn how to make a black hole in a suitable place. Material launched into a rapidly spinning black hole in just the right way can produce useful energy with 100% efficiency; 100% of the material's mass could be converted into useful work energy. Such

Figure 10.4. Sir Roger Penrose (1931–)

a civilization could load its trash into a rocket, fly the rocket near the black hole, and jettison the trash into the black hole. If done properly, the rocket would return with more kinetic energy than it left with, which could be easily converted into useful energy. This civilization could generate more energy from its trash than it could possibly need. Remember how much energy is contained in the mass of one penny? A black hole is an ideal dumpster, a bottomless pit effectively removed from our universe. No matter how much trash you throw into a black hole, none of it can ever return to our universe. You can throw in hazardous waste, toxic materials, anything you want to get rid of—none of it will ever return, and 100% of the discarded mass can be recovered as useful energy. It's a completely clean process; all the trash goes into the black hole, is recycled into elementary particles, and is gone forever. (We will discuss black holes further in later chapters.)

A black hole is the perfect energy generator.

RENEWABLE ENERGY

Let's turn next to two renewable energy sources: solar and wind. Both are ultimately fuel-based, driven by sunlight generated by nuclear fusion. While sunlight is not indefinitely renewable, our Sun will be a prodigious source of energy for billions of years—longer than Earth will continue to be habitable. Figure 10.5 compares leading near-term energy alternatives in terms of costs and carbon emissions, with the caveat that energy costs are highly volatile. Fission, solar, and wind have almost no carbon emissions. Some climate experts say carbon emissions should not exceed the dashed vertical line due to global warming concerns, but the exact number is strongly debated. Solar and fission costs are shown with current costs (right end of arrows) and costs their proponents claim can be achieved with large-scale deployment (left end of arrows).

It's clear why we use so much coal: it's cheap. But coal has by far the highest carbon emissions. "Clean coal", gasifying coal for higher combustion efficiency, and capturing and burying the released carbon, is substantially more expensive, but reduces carbon emissions to near-acceptable levels. Wind and solar both have high start-up costs that must be recovered over the life of the generating facility. Currently, the cost of wind energy is competitive, but solar is about 3 times more expensive than coal. Proponents claim all U.S. energy needs can be supplied from wind in the Dakotas and sun in Arizona. Distributing this energy to users across the country requires new systems with new technologies and substantial investment. Conventional AC power lines radiate excessively if they span over 500 miles. Underground, superconducting, DC power lines operating at –320 °F seem like the best alternative. Also, since wind and solar energy cannot be produced 24 hours per day, every day, they must be supported by energy storage systems based on new technologies or by back-up energy sources based on coal, gas, or nuclear.

In 2009, M. Jacobson and M. Delucchi reported that all the world's energy could be supplied by solar and wind as early as 2030 at a cost "on the order of $100 trillion", excluding energy transportation and environmental mitigation costs. Their plan requires 170,000 square miles of solar arrays, an area larger than California.

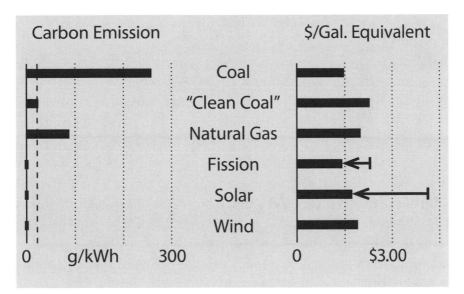

Figure 10.5. Charted above for several currently available energy options are: on the right, the cost of generating as much energy as contained in a gallon of gasoline; and on the left, the number of grams of carbon emitted per kWh of energy produced. Vertical dashed line at 22 g/kWh shows the recommended maximum carbon emission to ameliorate global warming.

LENR: LOW ENERGY NEUTRON REACTIONS

A remarkable theory by A. Widom and L. Larsen explains how nuclear-fusion-scale energy could be released in a novel way at normal temperatures and pressures by cleverly combining the electromagnetic, weak, and strong forces. They describe how to produce low-energy neutrons from collective effects. These neutrons could then be absorbed by nuclei without needing to overcome the repulsive electric force that necessitates the extreme temperatures in conventional fusion of positively charged nuclei. In opportune materials, repeated neutron absorption can produce new nuclei and release enormous amounts of energy. LENR could deliver the promised benefits of nuclear fusion, with a much simpler technology and without the extreme conditions that nuclear fusion requires. If LENR can be implemented on a commercial scale, it would be a tremendous boon to the planet and all its inhabitants.

11

Particles and Waves

Before Einstein, physicists thought particles and waves were two completely different phenomena. Particles appeared to be simple, but waves were clearly more complicated.

Waves are motion. They are not entities in and of themselves, but rather waves are the organized motion of a great many smaller entities. Figure 11.1 shows a typical wave.

Waves are characterized by *amplitude* and *wavelength*. Amplitude *A* is how high the wave goes up above and down below its average height. Wavelength *w* is the distance between crests. Waves move through space,

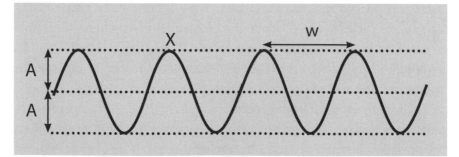

Figure 11.1. Waves never stand still, rather they are the organized motion of many small things, such as water molecules in an ocean wave. Waves have amplitude A and wavelength w. Point X is shown for reference.

Property	Particles	Waves
Localized	yes	no
Quantized	yes	no
Diffraction	no	yes
Interference	no	yes
Motion equation	$a=0$	$\Box A=0$
Medium	none	medium is required
Speed	any value	speed set by medium

Figure 11.2. Physicists' view of particles versus waves before Einstein

and change with time. If the wave in this figure is moving to the left, its entire shape moves together as if sliding left. If we concentrate on what the wave is doing at a specific point, such as point X in the figure, we would see the wave height going up and down as time passes; it oscillates between its maximum and minimum heights. As it goes from maximum to minimum and back to maximum again, the wave completes one full cycle. If the wave goes through 9 full cycles per second, its *frequency f* is 9 cycles/second, or 9 *Hertz*, which is usually written 9 Hz. A computer that executes 5 billion instruction cycles per second has a frequency of 5 gigahertz (5 GHz). The product of wavelength and frequency is always equal to the wave's velocity: *wf=v*.

Before Einstein, what physicists thought they knew about particles and waves is shown in figure 11.2.

They thought particles were like small baseballs, *localized* and *quantized*. Baseballs are always in definite locations and are easily counted. Waves are different; they are neither localized nor quantized. Sound waves spread to fill a room; outdoors they spread over vast areas, like sonic booms. Waves are measured by how much, not how many. We don't count sound waves; we measure the intensity of sound (loudness).

Intensity can have any value: it is **continuous**. The "amount" of particles is not continuous; it is quantized—we never see 2½ baseballs. The motion of particles is simple. Unless a force acts on a particle its acceleration is zero, *a=0*, and its velocity remains constant. As Newton said, "an object in motion stays in motion, an object at rest stays at rest, unless acted upon by a force." Waves have a much more complicated equation of motion [1].

Particles can move through air, through water, and also through empty space. Before Einstein, physicists thought particles could travel at any speed, given enough energy. Waves are different. Waves need something to travel through, a **medium**. In fact, in a wave it is the medium that moves—a wave is the organized motion of its medium. Ocean waves need water, and "waves" in a sports stadium need enthusiastic fans. Sound waves are the motion of air molecules. They can start with a car pushing air aside as it drives. The air is compressed, its pressure rises, and it pushes on neighboring air molecules that in turn push on their neighbors, etc., something like a cascade of dominos spreading outward in all directions. The speed of a wave is determined by the properties of the medium. Loud sounds travel at the same speed as soft sounds. Sound travels through air at the same speed regardless of the source, be it a person's voice, a car going 60 mph, or a jet plane. This is because sound is the motion of air, not the motion of a car or a jet. That is why we can say "the speed of sound" rather than "the speed of car sound."

Waves also **diffract** and **interfere**—strange things no self-respecting particle would ever do.

Diffraction is illustrated in figure 11.3. Here waves are moving from left to right; the white areas represent wave crests and the black areas represent troughs. When a wave hits a barrier with a small hole, part of the wave passes through the hole and **diffracts**—it spreads out across a wide range of directions. The greatest wave intensity goes straight ahead and broadens as it moves away from the barrier, becoming much broader than the hole it passes through. There are also side lobes of lesser intensity moving away from the hole and diverging from the central wave. These side lobes are separated from one another and from the central wave by dead zones, narrow strips where there is no wave intensity at all.

Particles do not diffract. If a group of particles hits a barrier which

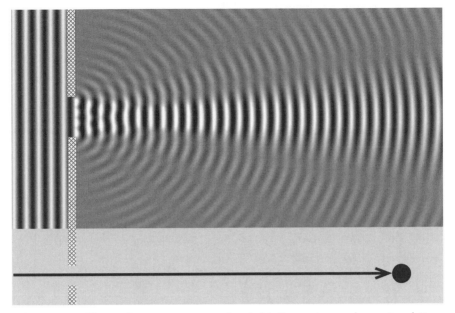

Figure 11.3. Above: A wave moves to the right; its crests are shown in white and its troughs in black. It diffracts after passing through a small hole in a barrier. Below: Particles don't diffract; they move in straight lines.

has a hole, those particles passing through the hole simply keep going in perfectly straight lines.

Wave interference, shown in figure 11.4, is also strange. Here waves from two sources spread out and *interfere* with one another. Again, the crests are white and the troughs are black. The waves reinforce one another where their crests coincide; this is *constructive interference*. Where one wave crests when the other troughs, they cancel one another; this is *destructive interference*. Again there are dead zones—narrow strips, where the waves cancel one another completely, resulting in zero intensity.

What would we see if we placed a piece of photographic film along the right side of figure 11.4? If only source A were turned on, it would completely expose the film. If only source B were turned on, it would also completely expose the film. But with both sources turned on, the film would have alternating bright and dark stripes called *interference fringes*. No light reaches the dark fringes because that is where the light waves

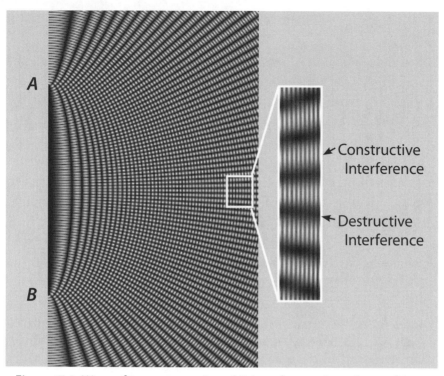

Figure 11.4. Waves from sources A and B interfere and produce a fringe pattern. White indicates high wave energy and black is low or zero wave energy. Interference is constructive where waves crest together and it is destructive where one wave crests when the other troughs.

interfere destructively. Each source prevents the other from illuminating places that either source would have exposed by itself. Having two ways to get there stops light from getting there at all!

By the end of the 19TH century, it was well known that light exhibits diffraction and interference, and Maxwell had shown that light obeys a wave equation. Everyone agreed light was a wave, and physicists generally rejected Newton's earlier idea that light was composed of particles.

Clearly, particles and waves were completely different.

At least, that is what everyone believed before Einstein explained the mystery of the photoelectric effect in 1905.

PHOTOELECTRIC EFFECT

When light hits a metal surface, electrons can be knocked out of the metal; this is called the *photoelectric effect*. It seems that electrons are ejected only when the energy of the light is high enough. That makes sense; electrons are attracted to the metal's positive nuclei, thus it takes energy to knock them away. But the devil is in the details. Color Plate 1 illustrates that the critical factor is the color of the light—its frequency— not the amount of light—its intensity. When high-frequency light, such as blue light, is used, electrons are ejected. If the intensity of blue light is reduced, the ejection rate drops, but it never drops to zero. When low-frequency light, such as red light, is used, no electrons are ejected even if the intensity of red light is very high. This is strange. High-intensity red light has more energy than low-intensity blue light. So what's wrong with red?

This makes no sense if light is a wave. However, Einstein realized it did make sense if light is composed of particles. A beam of light, he said, is not a *continuous* stream of energy, as we perceive a river to be a continuous stream of water. A beam of light really consists of a vast number of *discrete*, individual particles called *photons*.

When photons, even a vast number of them, hit a metal surface, an electron in the metal can be hit by only one photon at a time. An electron can be ejected only if a single photon has enough energy to knock it out. A thousand little nudges from low-energy photons will not do; it takes one good whack from one high-energy photon to eject an electron.

A photon's energy E, Einstein continued, is related to its frequency f according to his new equation $E=hf$, where h is a number called *Planck's constant*. Thus the frequency of the light must be high enough for one photon to have enough energy to eject one electron all by itself. High-frequency blue photons have enough energy; low-frequency red photons do not.

For explaining the photoelectric effect and discovering the particulate nature of light, Einstein received the 1921 Nobel Prize in Physics.

PARTICLE-WAVE DUALITY

Einstein realized light has both particle and wave properties—it is both a particle and a wave, two things everyone else believed were totally incompatible. Following Einstein's lead, we now know "particles" and "waves" are really labels for the two opposite ends of a continuous spectrum. "Particles" and "waves" are labels like "black" and "white." Everything in our universe is really a shade of gray. In the macro-world we live in, almost everything is either very nearly a particle or very nearly a wave; we usually don't notice the tinge of gray. But, in the micro-world, gray rules. Sometimes particle-waves are more particle-like, and sometimes more wave-like. But fundamentally, everything really is both.

Einstein introduced this remarkable idea, later called ***particle-wave duality***. It had enormous consequences in the development of Quantum Mechanics, and ultimately it led to conclusions Einstein could never accept.

NOTES

[1] The wave equation is not used in this book, but is shown below for your amusement:

Wave equation: $d^2A/dx^2 + d^2A/dy^2 + d^2A/dz^2 - d^2A/dt^2 = 0$

Shorthand: $\Box A = 0$

WHY FUND SCIENCE?

Answer: For love and money.

Love: Why do we devote so much time and money to music, literature, art, and sports? Because they enrich the human soul. They make life exciting and enjoyable. So does climbing a mountain to see what's on the other side. And so does discovering the mysteries of nature.

Money: Love is great but we all have to eat. We should invest in science because it's the engine that drives our technology-based society. Consider the country described below:

- The average life expectancy is 47 years.
- The average worker earns $300 per year.
- Only 8% of homes have a telephone.
- Cross-country phone calls cost $4 per minute.
- Only 14% of homes have a bathtub.
- There are no TVs, cell phones, or computers.
- There are only 144 miles of paved roads.
- Only 6% of the population graduates from high school.
- Only 10% of all its physicians have *any* college education.

Do you want to guess which third-world country this is? The answer is below. Economists estimate that 25% of our society's progress results from capital investment, e.g. buying a second mule to plow more land and grow more food. They estimate that 75% of our progress results from the advance of science and technology, e.g. inventing tractors to replace mules.

To improve all our lives, we must invest in science!

Answer: The United States of America in 1904.

12

Galileo and the Principle of Relativity

The Principle of Relativity was known long before Einstein.

Galileo Galilei, an Italian scientist of the late 16TH and early 17TH centuries, was the first to realize the necessity of testing scientific ideas with experiments and the first to emphasize the importance of Relativity in physical laws.

Before Galileo, science was dominated by the approach of the ancient Greeks: philosophical discourse. Scientific questions were decided by argument and by weight of authority. Aristotle was considered the ultimate authority on science.

The ancient Greeks contributed enormously to the advance of knowledge and culture, but they also clung to mistaken scientific ideas. They preferred to divine through discourse the ideals to which nature should conform, instead of observing actual natural phenomena. They believed everything had its own natural condition—heavy objects "should" fall, thus heavier objects "should" fall faster. Had they tested that idea, they might have scooped Galileo by 2000 years by discovering that all objects fall at the same rate. They were skeptical of physical observation that spiteful demigods might manipulate, and believed logic was the more certain path to understanding ultimate, and perhaps hidden, Platonic truths. Their approach dominated science for two millennia.

Galileo was the first to systematically apply mathematics and

Figure 12.1. Galileo Galilei (1564–1642), the father of modern science, used novel experimental techniques and developed instruments, such as the inclined plane, to more precisely study the motion of falling bodies. Dotted lines indicate the ball's position 1, 2, 3, 4, 5, and 6 seconds after its release.

experimentation to science. His approach may well have been inspired by his father, a famous lutenist, who discovered that the pitch of a lute string was proportional to the square root of its tension. Galileo said science "is written in this grand book, the universe … in the language of mathematics." He wanted to study this "book" and discover what really happens in nature, not what Aristotle thought should happen. How fast do objects fall, and do different objects fall at different rates? Galileo developed instruments and created special circumstances to probe more deeply and carefully (see figure 12.1). Without modern clocks, he devised clever techniques to quantify time, using his pulse to measure the period of pendulums. He rolled balls down inclined planes to slow their fall and allow more precise measurements of their motion.

Galileo discovered that when any object falls, the distance it travels increases with the square of time. He also discovered the principle of inertia—that objects move with constant velocities unless acted upon by an external force. His rejection of the blind acceptance of authority created a revolution that separated science from dogma.

> Galileo was the first modern scientist,
> and the inventor of experimental science.

One of his discoveries was the Principle of Relativity (we will update this when we discuss General Relativity):

> The Principle of Relativity
>
> The Laws of Nature appear the same
> to all observers moving with constant velocities.
> Absolute velocity has no physical meaning.
> Only relative velocities are meaningful.

Galileo observed that objects fall in the same way on a moving ship as on land. The modern version might be: if you spill your coffee while "dining" in an airplane, it falls straight down onto your lap even though the plane is flying at 600 mph. At first, it's not surprising the coffee falls straight down; everything does. But consider how this looks from outside the airplane: the coffee isn't just falling down, it's also moving forward at 600 mph along with everything else in the plane, even though nothing is pushing it forward as it falls. That would have surprised Aristotle.

What does the Principle of Relativity mean? Consider figure 12.2. Imagine two laboratories, each with a scientist and all the equipment they might desire. *Stop!* That's impossible; make that "a lot of equipment." Seal the labs so the scientists cannot see anything or measure anything outside their own lab. Then put each lab in its own airplane, with one

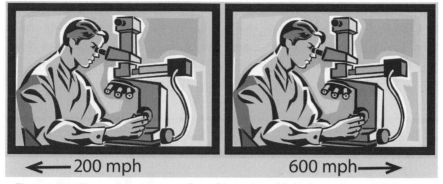

Figure 12.2. Two scientists are placed in isolated labs in separate airplanes. No measurement they can make will determine their airplanes' speeds.

flying west at 200 mph and the other flying east at 600 mph. The Principle of Relativity says there is no test or measurement the scientists can perform to determine their planes' velocities. Every test they make gives exactly the same results in both planes, as long as they cannot detect anything outside their labs. They could find their planes' speeds with **GPS**, but detecting an external signal is cheating. Relativity says all observers see the same physical phenomena, the same laws of nature, regardless of their own velocities, as long as their velocities are constant.

Absolute velocity is not detectable. But we can detect when an object's velocity changes or when two objects have different velocities. If one of the above airplanes suddenly stops (let's not ask why), we could measure the velocity change (deceleration) with a pendulum. The pendulum would not hang straight down, it would tilt toward the plane's nose, just as airline passengers have the sensation of being pulled toward the front of the plane when the pilot hits the brakes after landing. Velocity changes, such as 200 mph less than before, are relative and they are physically meaningful and measurable. So are velocity differences; if we observe an airplane from the ground, we can measure its speed relative to us.

It's a very good thing that the laws of nature don't depend on our "true" or "absolute" velocity, because we don't really know what that is. You might think that while you are sitting in a chair reading this book your velocity is zero, but is it? As shown in figure 12.3, as the Earth spins on its axis once daily, it carries us eastward at up to 1000 mph.

As Earth orbits the Sun, we're being carried along at 70,000 mph. As the Sun orbits the Milky Way galaxy, we are moving at 500,000 mph. And we're moving at 800,000 mph through the ***cosmic microwave background*** radiation, the first light of our universe that we will discuss in chapter 34. So, how fast are you really moving?

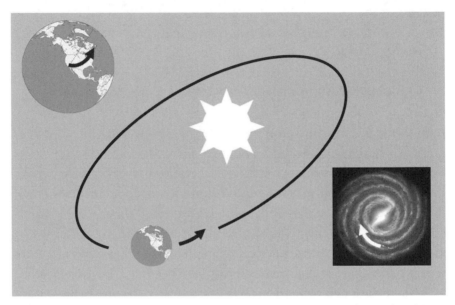

Figure 12.3. Earth's rotation carries us east at up to 1000 mph. Earth orbits the Sun at 70,000 mph. Our solar system orbits the Milky Way galaxy at 500,000 mph. How fast are we really moving?

An observer moving with constant velocity is said to be in an ***inertial frame***. Newton's laws of motion—an object at rest stays at rest, an object in motion stays in motion, etc.—are only valid in inertial frames. Special Relativity is also valid only in inertial frames. In inertial frames, physics is simpler, and who doesn't want simpler physics?

Earth's surface is generally considered to be an inertial frame, except in high precision applications. This is because we can generally ignore or compensate for Earth's gravity and rotation. An example of a non-inertial frame is a 78 rpm vinyl record (what music was recorded on before cassettes, CDs, and iPods). If we stand next to a record player (in

Earth's inertial frame), a ping-pong ball rolling across the record would appear to us to move in a straight line (absent friction). For us, the laws of physics are simple: no force means no acceleration, which means no change in the ball's speed or direction. But in the non-inertial frame of the rotating record, the ball's path is curved. Here the laws of physics are more complex: we must invent a fictitious force, a *Coriolis force*, to explain the ball's apparently curved trajectory. Whenever possible, it's better to avoid non-inertial frames.

ARE RELATIVITY AND MAXWELL COMPATIBLE?

If all this was known before Einstein, why is Relativity such a big deal? Since Newton's laws conform to Galileo's Principle of Relativity, there was no problem. At least not until 1861 when Maxwell published the theory of electromagnetism, and said light was an electromagnetic wave that moved at a fixed speed. That was a problem.

As discussed earlier, before 1905 physicists believed that all waves require a medium to travel through. No one knew what light's medium was, but they gave it a name, the *luminiferous ether*, or *ether* for short.

If light does need ether to travel through, ether must fill the entire universe because we can see light from very distant galaxies. Note that this situation is different from other waves, such as sound waves. Sound needs a medium, such as air, but since no one hears the sound of stars, no one claims the universe is filled with air (it isn't). Hence, light and ether pose a new and unique problem.

We must be moving through the ether if the universe is filled with it. People once thought Earth was the center of all that existed and everything revolved around us. Now we know that even our own galaxy is but a small part of a much vaster universe. We also know Earth is moving relative to everything else in the cosmos, thus it must also be moving relative to the ether and this means we should see light moving at different speeds in different directions.

Why? Consider the following analogy. If a rock is thrown into a pond, ripples spread across the water moving at the same speed in all directions.

If a rock is thrown into a river, a person on the bank sees the ripples moving faster going downriver and slower going upriver, because the current carries everything downriver. If a "river" of ether is flowing past Earth, light's speed must be higher going downriver than upriver.

But Maxwell's equations say the speed of light is a fixed number and is the same in all directions. Hence, Maxwell's equations would be correct in only one inertial frame, the frame in which the ether is not moving. We could then measure our absolute velocity, our velocity relative to the ether, contradicting the Principle of Relativity. Ether would make Maxwell's equations and Relativity incompatible.

Physicists set out to see if light's speed really did vary. Very precise and meticulous experiments measured its speed in different directions. They all found the same light speed in all directions, day or night, regardless of Earth's position orbiting the Sun. Experiments generally cannot prove two measured quantities are exactly equal; there are always some limitations to instrumental precision. The best that experiments can say is that two quantities are equal within a certain level of precision. If we measure two yardsticks, we might be able to say they are the same length to $1/16^{TH}$ of an inch. With better instruments, we might be able to say they are equal to $1/1000^{TH}$ of an inch, but we can never say they are *exactly* the same length. While exactitude is impossible, experimental physicists can sometimes achieve amazing levels of precision.

The most famous of these measurements was done in 1887 by American physicists Albert Michelson and Edward Morley. Michelson developed the first *interferometer*, whose descendants are still the gold standard of precise optical measurements. Michelson's interferometer, illustrated in figure 12.4, had two arms set perpendicular to one another. Each arm had a mirror at its far end. He sent light back and forth along each arm, and very precisely measured the difference between the travel times along the two paths. Michelson slowly rotated the entire interferometer keeping the arms perpendicular. If ether existed, the travel times would have changed as the direction of the light beams going through the ether changed. However, he found no changes in travel times to a precision of $1/40^{TH}$ of what was expected from Earth's motion—pretty convincing.

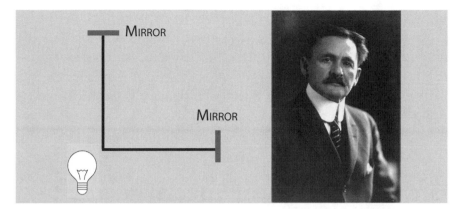

Figure 12.4. Albert Michelson (1852–1931) invented the interferometer to precisely measure changes in the speed of light along two arms set perpendicularly with mirrors at their ends. As the interferometer was slowly rotated, he found no speed changes, proving the speed of light is the same in every direction. He received the 1907 Nobel Prize in Physics for his outstanding experimental work.

Michelson was the first American to receive a Nobel Prize in science. Surprisingly, despite being honored for proving there was no "river" of ether flowing past us, Michelson continued to believe that ether really did exist. Long after Einstein's rejection of ether became widely accepted, Michelson struggled to understand how his great experiment had gone wrong. The truth is, the experiment wasn't wrong. Since 1887, this measurement has been redone a great many times, with ever greater precision. The latest experiments find light's speed is the same in all directions to a precision of 1 part in 10^{18}. If there is a change in the speed of light from one direction to another, that change would have to be less than one millionth of one millionth of one millionth of light's speed.

Does ether really exist? If it does exist, why can't we detect it?

13

Einstein's Theory of Special Relativity

Before Einstein, physicists were confident they understood distance, time, and mass. They believed any skilled person could measure distance with a ruler, tell time with a clock, weigh mass with a scale, and get the TRUE values. With his Theory of Special Relativity, Einstein proved that this was wrong—that there were no TRUE values—and he shook the very foundations of science.

Einstein developed two theories of Relativity: Special Relativity and General Relativity. Special Relativity, published in 1905, covers only special, simple situations, those in which there are no forces. It was followed ten years later by General Relativity, which extends Relativity to virtually all situations. General Relativity includes Einstein's theory of gravity and is the more complex theory. Let's start with Special Relativity.

Special Relativity supersedes Newtonian mechanics: his basic laws of motion and his concepts of space and time. It is not so much that Einstein's ideas contradict Newton's, rather Einstein's ideas address a broader range of circumstances. Newton's laws are still adequate for almost all day-to-day applications—bridge design, planetary motion, and billiards. But in more advanced applications, particularly when velocities approach the speed of light, nature deviates from Newton's laws. It is in those cases that we need the greater accuracy and deeper understanding of Special Relativity. If the speed of light were infinite, nothing else could approach its speed and there would be no need for Special Relativity.

In the previous chapter, we discussed the search for ether, the presumed medium of light, and how the existence of ether would create a conflict between Maxwell's equations and the Principle of Relativity. By the start of the 20TH century, the inability to find ether was identified by Poincaré as one of the "last few problems" in physics. Einstein wanted to solve this important problem. But he was also bothered by what he perceived as a peculiarity of the equations of electromagnetism—they give two solutions to the same problem.

Einstein's 1905 paper on Special Relativity is one of the most important in science. He began by explaining his concern about the electromagnetic equations. According to a "constant-field" equation, if a wire moves past a stationary magnet, as illustrated in figure 13.1, the unchanging *magnetic* field causes an electric current to flow in the wire. But if a magnet moves past a stationary wire, a "changing-field" equation says an *electric* field is created that causes a current to flow in the wire. Both equations yield the same current and most physicists were unconcerned that these equations provide two different ways to get the same answer.

But to Einstein, this was ugly, and he was sure nature wasn't ugly. Because absolute velocity has no

Figure 13.1. Einstein said, if a wire moves past a magnet, or if a magnet moves past a wire, the physics must be the same.

physical meaning, it cannot make any difference whether it's the wire or the magnet that moves. All that matters is their relative motion—their velocity relative to one another.

The Principle of Relativity requires that all observers see the same physical outcomes. Observers can disagree about specific measurements, as we will learn shortly, but they must agree about the occurrence of physical events—such as current flowing in a wire, objects colliding, or particles getting closer. Since natural phenomena depend only on relative velocities, Einstein believed that physics should follow nature and have a single equation that depends only on the relative velocity between wire and magnet.

Einstein resolved all these issues with his Theory of Special Relativity, based on these postulates:

> ### Einstein's Postulates of Special Relativity
>
> The Principle of Relativity applies to all physical laws.
> Maxwell's equations are valid in all inertial frames.
> Light travels without a medium; no ether is required.
> The speed of light is the same in all inertial frames.

Einstein said Relativity and Maxwell's equations are compatible after all: they are both valid for all physical laws and in all inertial frames. He said physicists had been mistaken: (1) to believe in ether; and (2) to inadequately understand how distance, mass, and time are actually measured. Einstein had thought much more carefully about this than anyone ever had before. Without ether, all observers must see light traveling at the same speed through empty space; we call that speed c.

Einstein did not prove his postulates of Special Relativity. He believed they were logical and "natural." He assumed these postulates were true and made predictions based on them. Innumerable experiments have confirmed the predictions of Special Relativity to extraordinary precision. Einstein's Theory of Special Relativity is the most extensively validated and most firmly accepted theory in science. Let's explore what Special Relativity tells us.

Consider the jet in figure 13.2, which is firing a laser directly forward.

Figure 13.2. A jet flying at ½c fires a laser directly forward. What is the laser beam's speed relative to the air? Is it 1c+½c=1½c? Einstein said the speed of light is always c, and he derived a new formula for adding velocities.

To the pilot, the laser beam is moving away from the jet at speed *c*; since a laser beam is light, it must move at the speed of light. This jet is really fast, flying at half the speed of light, ½c (your tax dollars at work). Question: how fast is the laser beam moving relative to the air? To help answer this question, consider a more familiar situation. If someone swam at 2 mph downstream in a river flowing at 5 mph, they would be moving 7 mph relative to the shore—one simply adds the two velocities. This makes it tempting to add *1c* and *½c* to get *1½c*; right? **Wrong!** Einstein said light's speed is always *c*. How do we add these two velocities and get a sum of *1c*? Einstein's answer was that due to the way we measure distance and time we need a new equation for adding velocities [1].

IT'S RELATIVE

Einstein said our measurements of distance, mass, and time are different from the pilot's measurements. Distance, mass, and time measurements are *relative*—they are not the same for all observers; they have different values in different reference frames, as we will see. In particular, when we observe a moving object, like the jet, it appears to us to be shorter, heavier, and its clock seems to run slower than an identical stationary object. The length, mass, and clock rate all change by the same factor that I denote by the symbol *g* [2], in the manner shown below:

	Some Measurements are *Relative*	
Example: Observe two identical jets with different velocities		
What We Observe	**Jet when Stopped**	**Jet when Moving**
Jet's Length	*d*	*d/g*
Jet's Mass	*m*	*m* × *g*
Jet's Clock Rate	*t*	*t/g*

Note that only distances along the direction of motion—lengths—seem shorter; widths and heights are not affected.

Why didn't anyone discover this before Einstein? One reason is that Einstein really was a genius. Another reason is that the impact of Special Relativity is very small at normal velocities, where g is almost 1, and is generally only important at velocities close to c, 671 million mph.

Figure 13.3 shows the impact of g. At $v=0$, $g=1$, and the formulas in the above box give us the normal values for a stationary object, as they must at $v=0$. At $1/10^{\text{TH}}$ of light's speed, 67 million mph, g increases by only ½%. But g grows rapidly above $v=0.9c$. The fastest charged particles I've dealt with had $v=0.999,999,999,7c$ and $g=40,000$. At $v=c$, g is infinity. This means at $v=c$, the jet's mass mg would be infinite. Thus an infinite amount of energy is required to accelerate the jet to the speed of light. Since nothing can have infinite energy, no object with mass can ever travel at the speed of light or faster. Conversely, particles that have zero mass, photons and gravitons, must travel at the speed of light or else they would have zero energy and would not exist.

WHAT TIME IS IT?

Newton assumed time was absolute and universal. He assumed time ran at its own intrinsic rate, unaffected by anything else, flowing at the same rate everywhere and always. Time was the same everywhere in the universe and could be kept by a single clock, perhaps Big Ben. Anyone

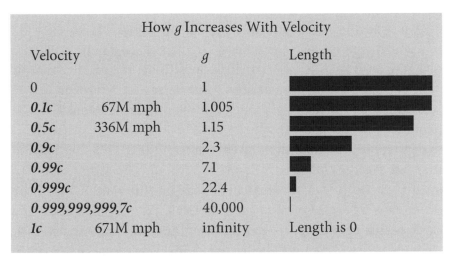

Velocity		g	Length
0		1	
0.1c	67M mph	1.005	
0.5c	336M mph	1.15	
0.9c		2.3	
0.99c		7.1	
0.999c		22.4	
0.999,999,999,7c		40,000	
1c	671M mph	infinity	Length is 0

Figure 13.3. The relativistic factor g grows as velocity approaches c, the speed of light. It would take an infinite amount of energy to boost an object with mass to the speed of light. This is why going faster than c is impossible.

could phone 1-800-4BIGBEN and a nice voice would say: "At the beep, the time everywhere in the universe will be..." Einstein showed that all this was wrong. Time is relative; there is no one, TRUE time. Time runs at different rates in different inertial frames; different clocks will inevitably measure different times.

Let's see why this is true. Time is our way of measuring how rapidly things change. If nothing ever changed, time would be a meaningless concept. Every clock counts the number of "times" something changes.

Let's try a ***thought experiment***. Einstein loved thought experiments; they were the only experiments he ever did. With thought experiments, Einstein kept his hands clean and avoided high voltage shocks, liquid nitrogen burns, and radiation (all the things that made me the man I am today). Thought experiments eliminate as many practical limitations as possible and focus on key principles. As Einstein said: "Make it as simple as possible, but not ***too*** simple."

So here we go. Imagine an ideal clock made of two perfectly reflective, exactly parallel mirrors with a single photon bouncing back and forth between them, as shown on the left side of figure 13.4. Each time the photon hits the top mirror, our clock will count one tick.

Now imagine a second identical clock moving rapidly to the right in a jet flying at *0.9c*. The jet's crew sees their clock standing still and operating exactly as we see our clock operating; they see their clock's photon going straight up and down. But we see everything in the jet moving rapidly, including the clock's mirrors and its photon. Thus the photon in the moving clock appears to us to have a longer path to travel to register one tick. Remember, every observer sees all photons move at the same speed. While our clock on the left counts 100 ticks, we see the moving clock on the right count fewer ticks. For a clock moving at *v=0.9c*, *g=2.3*, thus we see it count *100/2.3=43* ticks. We observe the moving clock measuring time passing more slowly than does our stationary clock.

Should the jet's crew get a better clock? That won't help. The Principle of Relativity requires all clocks moving at the same velocity to measure time at the same slower rate. If a second type of clock measured the "right" time, the difference between these two types of clocks could be used to measure the jet's absolute velocity, which Relativity says is impossible.

Everything in the moving frame of the jet seems to us to be running slower, not just clocks. We observe the crew's hearts beating slower, their brain waves waving slower, and their hair growing slower. Every biochemical process in their bodies appears to proceed slower, their lives proceed slower and last longer, and all at exactly the same slower rate.

In truth, it is time itself that appears to us to be slower in a moving frame; this is called ***time dilation***. We can use the clocks of figure 13.4 and simple geometry to find the formula for time dilation and g [3].

Relativistic effects are symmetric. The jet's crew would say their clock is fine, and it is our clock that's running slowly and it is we who are shorter and heavier. They observe all the same strange effects in our frame that we observe in theirs. Who is right? ***We both are!*** This seems impossible. But that is the way nature really works, and it all boils down to how measurements are really made. Because the speed of light is not infinite, moving people and stationary people simply do not make the same measurements.

There is no single right answer to "what time is it?" The answer really is different in different inertial frames—for people moving with different velocities.

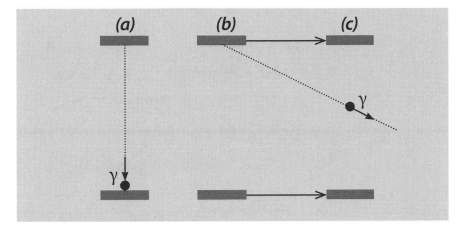

Figure 13.4. A stationary clock (a) has two mirrors (gray) with a photon (γ) bouncing between them. At time t=0, a second identical clock starts at (b) and moves to the right with velocity v=0.9c. Also at t=0, photons start moving down from the top mirror of each clock. When the left photon reaches its bottom mirror, the right photon and its mirrors are at (c). While the stationary clock counts 100 ticks, the moving clock counts only 43.

SPACETIME

Hermann Minkowski, Einstein's college math professor, was unimpressed with young Albert, calling him a "lazy dog." He may have reconsidered after the publication of Special Relativity. Minkowski realized Einstein's theory changed our understanding of the geometry of our universe. The three dimensions of space can no longer be considered distinct from time; space and time must be viewed as a combined four-dimensional entity that Minkowski named *spacetime*. In four-dimensional spacetime, points are called *events* and four quantities are required to specify an event, such as latitude, longitude, altitude, and time. Time measurements and normal three-dimensional distance measurements are relative (different for different observers), but the distance between two events in four-dimensional spacetime is *invariant* (the same for all observers).

Spacetime can be thought of as salami, as in figure 13.5. In our inertial frame, a normal slice perpendicular to the long axis is a slice of constant time; we observe every event on that slice occurring at the same time,

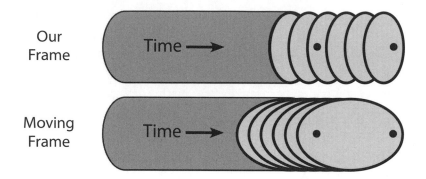

Figure 13.5. Time and two dimensions of space are shown here as a salami. Above: we slice spacetime perpendicular to our time axis; for us, every event on a slice occurs at the same time. Below: we see spacetime sliced askew in a moving frame, mixing time and space. Black dots are events that we see at different times, but which appear simultaneous in the moving frame.

to us they are simultaneous. The constant time slices of moving observers seem askew to us; events they deem simultaneous may appear to us to occur at different times. We see their time and their space mixed together.

SPACE TRAVEL AND THE TWIN PARADOX

Einstein's Special Relativity both constrains and enables distant space travel. With c as a universal speed limit, we can never send a rocket to another galaxy and have it return in our lifetime; such a voyage would take at least 300,000 Earth-years. Even at nearly the speed of light, we could only reach about 1000 stars and a billionth of our galaxy in a round trip lasting less than 100 Earth-years. But time dilation allows astronauts to reach more distant destinations within *their* lifetimes. If astronauts traveled at $v=0.999c$, earthlings would observe them living 22.4 times longer than normal. Their lives would have the same content as ours, the same number of heart beats, the same number of thoughts, etc. They would simply appear to live 22.4 times slower. If they lived 80 years in their frame, they could reach a destination 1800 light-years away, and still be alive 17 centuries after their siblings on Earth had died.

The twin paradox is a famous puzzle of Special Relativity that may amuse you. One twin rockets off to the Delta Quadrant and returns to Earth. Citing Special Relativity, each twin claims he was stationary and it was the other twin who moved and is therefore younger. Who is right?

If the rocket had never returned to Earth, they would both be right. But to return, the rocket had to decelerate near the distant star, turn, accelerate, and finally decelerate again to stop here—changing velocity at least three times. Thus the astronaut twin was not in an inertial frame during his whole trip and he cannot invoke Special Relativity. The Earth-bound twin is older and wiser.

 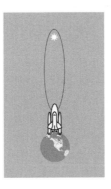 

Figure 13.6. Question: Which twin is younger?   Answer: Alber is.

NOTES

[1] The sum of collinear velocities u and v is $(u+v)/(1+uv/c^2)$. For the jet firing the laser beam: $u=1c$ and $v=0.5c$, so $u+v=(1.5c)/(1+0.5)=1c$.

[2] Standard physics notation uses γ, the Greek letter *gamma*, the same symbol that is used for photons. I use g to avoid confusion.

[3] The result is $g^2=1/(1-v^2/c^2)$. Note that all relativistic equations reduce to Newton's equations if the velocities are very small compared to c; the sum of collinear velocities [1] becomes $u+v$, and g becomes 1.

14

Einstein and Light

Light is energy without mass.

Einstein radically changed our understanding of light by explaining that light is both a particle and a wave, and that, unlike all other waves, light does not need a medium to travel through—light can travel through empty space. What exactly is light?

Einstein discovered that light consists of individual particle-like entities called *photons*. Each photon is a combination of electric and magnetic fields that oscillate like waves as light moves, as shown in figure 14.1. The orientations of the electric and magnetic fields are perpendicular to one another and both are perpendicular to the velocity. In the figure, light moves to the right. The electric field oscillates up and down within the page, while the magnetic field oscillates in and out of the page. Only the two fields wave; no material substance or other medium moves.

The key characteristics of photons are wavelength w, the distance traveled in one full cycle, and frequency f, the number of times per second the electromagnetic fields oscillate through a full cycle. Frequency determines a photon's energy E, according to Einstein's equation $E=hf$, with h being *Planck's constant*. For any wave, the product of its wavelength and its frequency equals the wave's velocity $wf=v$. For light moving through empty space $wf=c$.

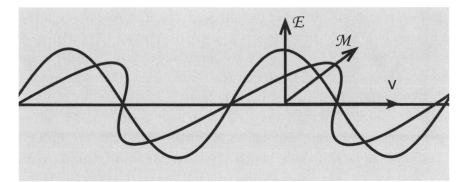

Figure 14.1. Light consists of oscillating electric $\mathcal{E}$ and magnetic $\mathcal{M}$ fields. No medium is required and no material substance moves. $\mathcal{E}$ oscillates up and down and $\mathcal{M}$ oscillates in and out of the page. The electric and magnetic fields are perpendicular to one another, and both are perpendicular to the direction of light's velocity v.

LIGHT'S SPECTRUM

Photons are categorized according to wavelength, or equivalently frequency, as shown in Color Plate 2. Photons with the shortest wavelengths, highest frequencies, and highest energies are called *gamma rays*. Descending the energy scale through the rest of light's spectrum are *x-rays*, *ultraviolet*, visible light, *infrared*, *microwave*, and *radio waves*. Within the narrow band of frequencies our eyes can see—visible light—the colors from highest frequency to lowest are: violet, blue, green, yellow, orange, and red. Other colors are composites of these.

Wavelength and frequency are not intrinsic properties of photons; they depend on an observer's inertial frame. As discussed in chapter 34, the cosmic microwave background (CMB) radiation is light that fills the universe. We observe it in the microwave portion of the spectrum, with a wavelength of 2 millimeters and a frequency of 160 GHz. But to particles moving at $v=0.999,999,999,7c$, CMB photons coming head-on would be *blueshifted*, as discussed in chapter 32. In that reference frame, the photons' frequency and energy is increased 40,000-fold and their wavelength is decreased 40,000-fold; they appear not as microwaves but as extreme ultraviolet radiation. But, these are the same photons in both frames.

They seem radically different only because the state of motion of the observers is radically different. Some people use the word "light" to refer to visible light only. I use "light" to refer to any electromagnetic radiation, because all photons are intrinsically identical. Differences among photons are only in the eye of the beholder.

Is the speed of light always c? It is when light moves through empty space, a perfect vacuum. But light slows down when it travels through glass, water, or even air. This is because photons are absorbed and reemitted by atoms in these materials, introducing delays. If the speed of light in a material is reduced to c/N, that material is said to have an ***index of refraction*** of N. For air $N=1.0003$, thus air slows light only slightly to $0.9997c$. For glass $N=1.52$ and for water $N=1.33$. For a girl's best friend, $N=2.42$ (one reason diamonds are dazzling).

As discussed in the last chapter, Einstein said c, the speed of light in empty space, was the highest possible speed anything can attain. That highest possible speed is not attainable for anything with mass, as that would require an infinite amount of energy, which is impossible. Particles without mass, photons and gravitons, always travel at the speed of light (in empty space), no more and no less; at any other speed their energy would be zero and they would not exist.

Are we sure the mass of photons is really zero? If it isn't, they cannot travel at the speed of light. This would not negate Special Relativity, but it would require a new name for the highest possible speed, perhaps e, the Einstein speed. It seems we are spared this confusion. The latest measurements of the photon's mass are consistent with zero and show its mass cannot be more than 10^{-24} times the electron's mass. The electron's mass is the smallest well-known mass (neutrino masses are smaller but are not well-measured). That means the photon's mass, if it isn't zero, is no more than a millionth of a millionth of a millionth of a millionth of the lightest known mass. Zero looks like a good bet.

NO NOBEL FOR RELATIVITY

For discovering the particulate nature of light, Einstein received the 1921 Nobel Prize in Physics.

There is an interesting story about this. In 1921, the Royal Swedish Academy of Sciences, which selects Nobel Prize recipients, could not agree on who should receive the physics prize, and they elected to defer the award for a year. In 1922, they gave the 1922 prize to Niels Bohr and the 1921 prize to Albert Einstein. When they notified Einstein that he was receiving the Nobel Prize, they pointedly informed him that it was for his solution of the photoelectric effect and *not* for his theories of Relativity. Furthermore, they said he was *not* to mention Relativity in his acceptance speech because they regarded his theories as far too speculative. Never one to kowtow to authority, Einstein thanked them but said he would not be attending the award ceremony due to "prior commitments."

The King of Sweden was displeased with these developments, and arranged a state dinner in Einstein's honor, requesting that he give a speech explaining Relativity. Einstein was delighted to accept.

The Royal Swedish Academy of Sciences thus rejected the greatest scientific discovery in two centuries. In 1912, however, they had given Swede Gustav Dalén the Nobel Prize in Physics for inventing a timing device to control lights on buoys. Go figure.

15

Al Makes Mom Proud

In 1905, Einstein published five spectacular papers that revolutionized physics, and answered the key questions of the day. Many say that 1905 was Einstein's Miracle Year. Actually his five papers were published within 7 months—he didn't even need the full year.

Einstein forever changed our understanding of:

Energy

Light

Mass

Space

Time

Figure 15.1. Einstein contemplating nature

Einstein thought his discoveries would set the world of physics on fire, and he eagerly waited for universities to beat a path to his door with glorious job offers. He waited... and waited... and waited. Nothing! Good thing he didn't quit his day job.

It took a very long time before the importance of his work was recognized. Not only were his conclusions revolutionary, but so were his analytical methods. Most physicists had trouble understanding what he wrote, let alone deciding if they believed any of it. And who was this Einstein fellow anyway? He was not a university professor, did not have a Ph.D., and did not work at an established physics institution. The real miracle of 1905 may be that Einstein's papers were published at all.

Gradually, important physicists, particularly Max Planck, began to see value in Einstein's ideas, and eventually some contacted him. Some wrote to point out "errors" in his papers (Einstein did not agree). Some asked questions, and some sought to discuss his discoveries. But still no job offers.

In 1907, in frustration, Einstein applied for a job as a high school teacher. But he was deemed under-qualified and wasn't even offered an interview. Finally in 1909, four years after his Miracle Year, and after spending seven years as a patent clerk, Einstein was offered his first academic position, a junior professorship at the University of Zurich. At the age of 30, he finally achieved his career goal: a university position as a theoretical physicist.

For the next four years, Einstein moved from university to university, slowing climbing up the academic ladder. Then in 1913, with Max Planck's enthusiastic support, Einstein received a spectacular job offer in three parts: (1) full professor at the University of Berlin, the world's leading physics institution, with the promise that he would have no teaching duties; (2) director of a new physics institute, with the promise that he would have no administrative duties; and (3) appointment to the prestigious Prussian Academy of Science. The multiple positions came with multiple salaries. With no teaching or administrative duties, he could spend all his time on his own research. He could pursue his own dreams, accountable to no one. It is hard to imagine a more perfect position for Einstein, or indeed for any scientist.

CAN PHYSICS BE BEAUTIFUL?

In 1915, Einstein published his greatest contribution to science, the Theory of General Relativity that supersedes Newton's theory of gravity. As British physicist Jonathan Allday said, General Relativity is "widely acknowledged as one of the most beautiful creations of the human mind."

Can a physics theory really be beautiful? Actually, physicists place great importance on the beauty of their theories. Like Einstein, most physicists believe nature is fundamentally harmonious, simple, and beautiful. Physicists believe an ugly theory must not properly portray nature. Experimental confirmation is the ultimate test of all scientific theories, but while awaiting definitive tests, beauty is considered an essential measure of a theory's merit. Almost every physicist considers General Relativity to be our most beautiful theory. Paraphrasing Robert Crease's description of beautiful experiments, General Relativity elegantly resolves deep mysteries of gravity, energy, space, and time; it reveals profound truths and transforms our understanding of the universe; and its conceptual simplicity, clarity, and power are joyful, surprising, and satisfying.

> Beautiful equations are the poetry of physics.

One of General Relativity's predictions is that starlight bends in the gravitational field of the Sun, or more precisely that light follows a curved path in a spacetime curved by the Sun. British astronomer Sir Arthur Eddington led a major expedition to test this bold prediction. At a meeting of the Royal Society in 1919, Eddington announced his findings confirming Einstein's theory. Reporters asked Eddington if it were true that he was one of only three physicists in the world who understood Einstein's theory. Eddington hesitated and then replied, "I'm just wondering who the third might be."

Major newspapers around the world ran front-page articles announcing the confirmation of General Relativity. They proclaimed the dawn of a new age of science, with Einstein as its undisputed prophet. Almost overnight, Einstein became a world celebrity, the Elvis Presley of physics, more famous than any scientist had ever been before or has been ever since. He had achieved unprecedented fame and admiration, both from physicists and from the public. At the age of 43, Einstein received the highest honor in physics, the Nobel Prize.

EINSTEIN'S ACHIEVEMENTS

Einstein's achievements are unparalleled:
- Solved mystery of Brownian motion
- Firmly established existence of atoms
- Discovered light is composed of particles
- Discovered equivalence of mass and energy
- Introduced concept of particle-wave duality
- Special Relativity—a new theory of mechanics
- Established theoretical basis for laser development
- General Relativity—a new theory of gravity, space, and time

Max Planck said there was hardly a problem in modern physics "to which Einstein has not made a remarkable contribution."

I would hazard to say that every research physicist relies on at least one of Einstein's discoveries every day.

Time Magazine chose Einstein as their *Person of the Century*, calling him the "greatest mind and paramount icon of our age."

> "Einstein was a giant.
> His head was in the clouds, but his feet were on the ground.
> Those of us who are not so tall have to choose!"
> — Richard Feynman, Nobel 1965

16

Einstein and Quantum Mechanics

Although Einstein was a major contributor to the theory of Quantum Mechanics, he was also its most formidable and dedicated adversary. Einstein ultimately rejected Quantum Mechanics because it embraces uncertainty—an embrace that has its roots in particle-wave duality. Einstein initiated the concept of particle-wave duality and thereby led Quantum Mechanics toward the eventual realization that nature is uncertain. Yet, it was a conclusion he refused to accept. Einstein's many important contributions to Quantum Mechanics include:

- Discovering light is composed of particles—photons.
- Introducing the principle of particle-wave duality.
- Developing, with Bose, a quantum theory of identical particles.
- Establishing the theoretical basis for the development of lasers.

Quantum Mechanics is our theory of the very small—of things much smaller than humans—molecules, atoms, and particles. Because the micro-world is vastly smaller than what we can directly perceive, humans evolved without any experience of it. Our innate sense of what is "natural" serves us well in the macro-world, but fails us completely in the micro-world. The strangeness of the world of Quantum Mechanics is illustrated by statements by two outstanding physicists who received Nobel Prizes for developing this theory:

> "If Quantum Mechanics hasn't profoundly
> shocked you, you haven't understood it yet."
> — Niels Bohr, Nobel 1922
>
> "I think that I can safely say that
> no one understands Quantum Mechanics"
> — Richard Feynman, Nobel 1965

The equations and procedures of Quantum Mechanics are now well known, but exactly why the micro-world behaves as it does remains a mystery. Two key principles of Quantum Mechanics are:

Quantization: Some properties are not continuous at small scales.
Particle-Wave Duality: Everything is both a particle and a wave.

QUANTIZATION

Quantization is not a totally strange idea; for example, money is *quantized*. In the U.S., the amount of money in any transaction is always an integer multiple of 1¢; we can say 1¢ is the *quantum* of U.S. currency.

Consider a physical example: staircases are quantized; ramps are not. On a ramp, one can be at any elevation. On a staircase, one can only be at elevations corresponding to the height of the steps. One's elevation can never be halfway between the height of the second and third steps, simply because there's no step there. This is illustrated in figure 16.1.

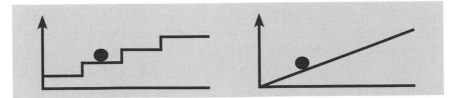

Figure 16.1. At a microscopic scale, properties such as energy are quantized like height on a staircase, instead of continuous like height on a ramp.

On a staircase, elevation changes substantially and suddenly from one step to the next. To contrast the macro-world with the micro-world, a planet can orbit the Sun at any distance: its energy is continuous (like height on a ramp). But, an electron in an atom can have only certain orbits: its energy is quantized (like height on a staircase), as we will discuss.

SOLVING THE "ULTRAVIOLET CATASTROPHE"

Quantum Mechanics began in 1900, when German physicist Max Planck solved the "ultraviolet catastrophe", which was the failure of Thermodynamics to account for light emitted by hot objects. Developed in the 19TH century, Thermodynamics very effectively explained the physics of heat, one of the many forms of energy (see chapters 2 and 9). Thermodynamics said that hot objects emit energy in the form of light. A grill gets "red hot" and some stars are "white hot." But the theory went further and said hot objects, such as lamps, emit an unlimited amount of high-frequency (ultraviolet) light. In fact, the theory said lamps emit an infinite amount of energy. But that's crazy: if lamps emitted infinite energy, they would cremate the universe. Obviously, this was wrong. While it wasn't a catastrophe on the scale of world war, physicists were unhappy that their nice theory was clearly flawed.

Planck suggested this "catastrophe" would be averted if the emitted light was quantized, like steps in a staircase. He said the size of one step is hf, where f is the frequency of light and h is a number we now call **Planck's constant**. Planck's idea made the total emitted energy finite, which "solved" the problem. Just as c sets the scale where Special Relativity becomes important, h sets the scale where Quantum Mechanics becomes important. If h were zero, the staircase steps would disappear and leave only a ramp, there would be no Quantum Mechanics, and the micro-world would behave just as Newton would have expected.

Planck was happy with this result and ultimately received the 1918 Nobel Prize in Physics. But he never thought his solution was anything more than a mathematical procedure, a clever trick. Einstein saw much more: he saw an important underlying physical reality that explained both

Planck's idea and the photoelectric effect that we discussed in chapter 11. Einstein realized that light is quantized: light consists of individual particles called photons. He saw that photons fit in perfectly with Planck's idea: *hf* is the energy of a single photon. Since the number of photons must be an integer (the number of particles is always an integer), the emitted energy must always be an integral multiple of *hf*.

One might think Planck would have been delighted that someone provided a physical rationale for his idea. He was not. Planck strongly supported Einstein; he was the first important physicist to promote Einstein's theories, he was instrumental in getting Einstein a prestigious job, and he forcefully recommended Einstein for the Nobel Prize. Nonetheless, Planck often said Einstein had gone too far with this notion of photons and insisted his own idea was just "a purely formal assumption."

PARTICLE-WAVE DUALITY

Einstein had opened Pandora's Box by discovering that particles and waves were not entirely separate phenomena. This launched the principle of particle-wave duality. Einstein could only watch as Quantum Mechanics expanded on this principle and moved farther and farther from his vision of nature.

French physicist Louis de Broglie extended Einstein's idea of particle-wave duality. He said that because waves have particle properties, as Einstein demonstrated with light, particles should have wave properties. In particular, they should have a wavelength *w* related to their energy and mass [1]. The higher a particle's energy, the shorter is its wavelength. For this contribution, de Broglie received the 1929 Nobel Prize in Physics.

The fact that particles have wavelengths is the primary reason for the strange behavior of the micro-world. These wavelengths are millions of times smaller than the distances we normally deal with in the macro-world, hence the effects of particle wavelengths are negligible in our macro-world. Only when we probe distances comparable to particle wavelengths do they become important; this is when we enter the micro-world of Quantum Mechanics.

PARTICLE WAVELENGTHS MAKE ATOMS STABLE

With de Broglie's theory of particle wavelengths, Quantum Mechanics is able to explain why atoms don't immediately collapse, which is very important to understand. Negatively-charged electrons "want" to be as close to the positively-charged nucleus as possible. Why don't they quickly spiral into the nucleus, perhaps releasing energy by emitting photons? If they did, the atoms we know would not exist and this book would end here. Thankfully this does not happen, as Danish physicist Niels Bohr explained with Quantum Mechanics.

Bohr said electrons cannot be in just any orbit around the nucleus. For an electron's wave to match evenly all the way around its orbit, the circumference C of the orbit must be a whole number of wavelengths w: $nw=C$, with n an integer such as 1, 2, etc. See figure 16.2. For each value of n there is an orbit with a specific electron energy: En. The smaller n is, the closer the orbit is to the nucleus, and the lower is the electron's energy. Electrons cannot spiral in closer than the $n=1$ orbit because n, the number of wavelengths, cannot be less than 1. Because the number of electrons in each orbit is limited, they stack up in an orderly progression starting with $n=1$ and going out from there (more in chapter 17).

When an electron moves from one orbit to a closer one, its energy decreases. Figure 16.2 shows an example of an electron dropping from $E2$ to $E1$. To conserve energy, it emits a photon with energy, $E=E2-E1$. For an electron to move to a higher orbit, say from $n=3$ to $n=4$, it must absorb a photon whose energy is $E=E4-E3$. These emission and absorption energies comprise a unique set of frequencies, a *spectrum*, for each type of atom, a unique fingerprint. Color Plate 3 shows the spectra of: hydrogen H; mercury Hg; and neon Ne. Clearly the spectrum of each element is quite different.

In 1842, French philosopher August Comte said that there were some things mankind would never know, such as the chemical composition of the stars. At least regarding the stars, Comte was wrong. Knowing the fingerprint of each element, we can analyze starlight and determine how much of each element is in each star. We can obtain precise, long-distance, chemical analyses of most celestial bodies.

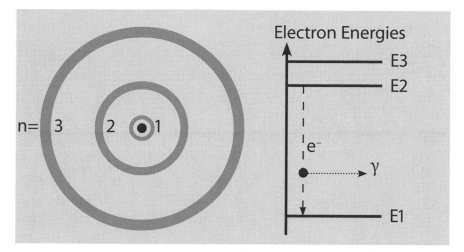

Figure 16.2. Left: Niels Bohr said that electrons' wavelengths restrict them to only certain orbits (gray circles labeled *n=1, n=2, n=3*, etc.) around a nucleus (central black dot). Right: Electron orbits have specific energies unique to each type of atom. If an electron (e⁻) drops from orbit 2 to orbit 1, it conserves energy by emitting a photon (γ) with energy *E=E2-E1*.

WAVE PACKETS

If particles are in some sense wave-like, why do particles appear localized while waves are not? Actually, pure waves aren't localized, but *wave packets* are. In a pure wave, everything waves at a single frequency, making the wave completely non-localized, as shown by the five examples on the left side of figure 16.3. Pure waves have no beginning nor end; they spread everywhere—they aren't localized at all. Adding together pure waves of various frequencies creates a wave packet, with the waves interfering as waves are wont to do. The waves all crest in the center of the packet, making the packet's energy largest in the center. The farther away from the center one looks, the more the waves cancel one another, reducing the packet's energy. Done properly, this creates a wave packet that is somewhat localized, within a distance *dx*, as shown on the right side of figure 16.3.

But, you don't get something for nothing. To get a highly localized wave packet you must combine waves with a broad range of frequencies *df*.

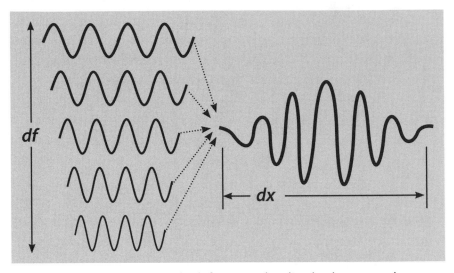

Figure 16.3. Pure waves on the left are not localized—they extend everywhere. Adding pure waves with different frequencies creates a wave packet, shown on the right, that is somewhat localized. The larger the range of frequencies *df* the shorter the wave packet *dx*. The smaller is *df* the larger is *dx*. It's not possible to have both a small *df* and a small *dx*. It's a trade-off.

To localize the wave packet within a small *dx*, one needs a large *df*. The smaller *dx* is the larger *df* must be; the smaller *df* is, the larger *dx* is—*dx* and *df* are inversely proportional.

QUANTUM UNCERTAINTY

Where is the particle within the wave packet? Somewhere within *dx*, but it's impossible to know exactly where within *dx* [2]. What is its frequency (energy)? It is within *df*, but it's impossible to know exactly where within *df*. It can be shown mathematically that the product *dx* × *df* cannot be less than a certain minimum value. This a limit that cannot be overcome. It isn't possible to have both a completely certain position (*dx=0*) and a completely certain frequency (*df=0*). We can trade uncertainties, but we cannot eliminate them entirely.

Werner Heisenberg was certain about this uncertainty and made it a principle (guess what it's called). He said there are inherent uncertainties in the properties of particles in the micro-world, not because our instruments are inadequate, but because these uncertainties are fundamental to the wave nature of particles. His principle is [3]:

Heisenberg's Uncertainty Principle

$$dt \times dE > h/2\pi$$

This equation merits explanation. Since a wave packet is a combination of different frequencies, as in figure 16.3, it is impossible to accurately determine a particle's energy by examining only a small portion of its wave packet. Measuring a larger portion of the wave packet reduces the energy uncertainty but requires letting more of the wave packet move past some instrument, which takes time. Thus there is a trade-off: the measurement time dt versus the energy uncertainty dE. One can become smaller only by making the other larger.

An interesting aspect of particle wavelengths relates to the "size" of elementary particles. Their intrinsic size is extremely small, probably as close to zero as anything can be [4]. Experiments show that the intrinsic size of elementary particles is no more than a billionth of the size of an atom. Quantum Mechanics says particles have zero size but cannot be localized more precisely than their wavelengths. The higher a particle's energy, the shorter its wavelength, and the more localized it becomes. With sufficient energy, elementary particles can be squeezed seemingly without limit. There seems to be nothing "solid" inside them.

THE EINSTEIN-BOHR DEBATE

Einstein **hated** all this. He believed, as did Newton, that the harmony and beauty of nature requires certainty. Both believed that if we knew the exact state of every particle at one moment, we could use the laws of physics to compute the exact state of every particle at all future times.

They believed the universe was completely deterministic. But, if we cannot know exactly where particles are and what their energies are, we cannot know exactly what will happen in the future. Einstein believed nature knew what would happen, even if Quantum Mechanics did not. He believed, therefore, that Quantum Mechanics was only an interim step on the path to a better theory, one without uncertainty.

Einstein conceived many thought experiments challenging Quantum Mechanics and its uncertainty; one is shown in figure 16.4. Here a clock mechanism opens a shutter *S* for a brief instant *dt*. The box contains a photon bouncing between mirrors, and during *dt* the photon exits the box through the open shutter. The photon's energy *E* is determined by the decrease in the clock box's mass, as measured on the scale *W*—its departure reduces the box's mass by E/c^2. Einstein said the box could be weighed very accurately, after the photon escaped, allowing the uncertainty in its energy *dE* to be extremely small. He said *dt* could also be extremely small. In fact, he said, there was no limit to how small either could be. Therefore $dt \times dE$ could be less than Heisenberg allowed. This, Einstein claimed, proved that Quantum Mechanics was wrong!

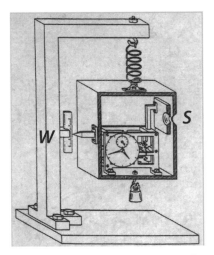

Figure 16.4. Einstein designed this clock box to disprove the Uncertainty Principle, but Bohr showed Einstein was wrong.

Bohr was Einstein's good friend as well as his principal adversary when it came to debating quantum theory. Bohr emerged as the leading proponent of the most commonly accepted view of Quantum Mechanics, the Copenhagen Interpretation.

At first, Bohr and the other "mechanics" of the quantum crowd had no answer to Einstein's challenge. But after many hours and a migraine, Bohr declared Einstein was wrong because he hadn't accounted for one of his own theories. Quantum Mechanics says that trying to precisely weigh the box creates an uncertainty in the box's

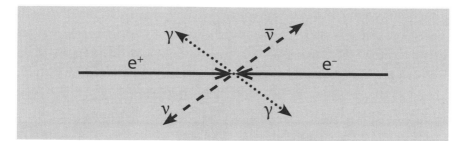

Figure 16.5. Particle interactions can have many possible outcomes. Here an electron (e⁻) and an antielectron (e⁺) collide and annihilate. They can produce a pair of photons (γ), or a neutrino-antineutrino (νν̄) pair, or other possibilities. Quantum Mechanics predicts the probability of each possible outcome, but cannot predict the outcome of any individual collision.

elevation in the gravitational field, and that, in turn, leads to an uncertainty in the measurement of time according to Einstein's Theory of General Relativity, as we will discuss in chapter 21. Thus Bohr concluded, *dt* could not be as small as Einstein wished because of Einstein's theory of gravity. (Hoisted on his own petard!)

The wave nature of particles prevents us from knowing their exact position and velocity. Therefore, when particles interact, we can never be sure what the outcome will be. Quantum Mechanics does allow us to exactly compute the probability of every possible outcome. For example, we can arrange 100 head-on collisions of an electron and an antielectron, as illustrated in figure 16.5. Quantum Mechanics predicts the percentage of collisions that produce photon pairs, the percentage that produce neutrino pairs, etc. However, it says that it is impossible to predict with certainty the outcome of the 101ST collision. Quantum Mechanics says the outcome of any specific interaction can never be predicted; it will always be a matter of chance.

Einstein hated this as well, remarking: "God does not play with dice." He believed the outcome of every interaction was pre-determinable and since Quantum Mechanics predicts only probabilities, it must be an incomplete theory. Bohr believed we must not cling to our ideas of how things "should be" and must instead accept nature as it is, telling Einstein: "Don't tell God what to do with His dice."

Figure 16.6. Albert Einstein (1879–1955) and Niels Bohr (right) (1885–1962) were lifelong friends who forever disagreed about the uncertainty of Quantum Mechanics and the nature of physical reality.

Ultimately, Einstein lost every battle against Quantum Mechanics and its strange view of reality in the micro-world. He never conceded defeat, but as the victories of Quantum Mechanics mounted, its proponents gained ever greater confidence and respect. In the end, Einstein's attacks helped them make Quantum Mechanics bullet-proof.

As time passed, Einstein's objections to Quantum Mechanics were supported by fewer and fewer physicists.

The man who had revolutionized physics as a bohemian outsider, never accepted the next revolution in physics when he was the ultimate insider.

Physicists' admiration for Einstein personally and for his great accomplishments never waned, but the mantle of leadership in physics passed to others.

NOTES

[1] More precisely: $w=h/mvg$, where w is the particle's wavelength, m is its mass, v is its velocity, h is Planck's constant, and g is the relativistic factor, which depends on v.

[2] Where is an electron within the electron "cloud" around a nucleus? Quantum Mechanics says it is everywhere simultaneously because its wavelength is that large. Are you shocked? Me too.

[3] Three other uncertainty equations (that are not used in this book) relate the uncertainty in a particle's position to the uncertainty in its momentum p, $p=mvg$. Momentum, like velocity, has directionality; the portion in the x direction is denoted p_x. The uncertainties in each direction are independent:

$$dx \times dp_x > h/2\pi$$
$$dy \times dp_y > h/2\pi$$
$$dz \times dp_z > h/2\pi.$$

[4] That there may be a lower limit to how small something can be in our physical universe is discussed in chapters 25 and 28.

17

Quantum Mechanics after Einstein

Relativity and Quantum Mechanics are the two great pillars of 20TH century physics. Both have been extensively tested with extraordinary precision, and none of the predictions of either theory has ever been falsified. For example, the value that Quantum Mechanics predicts for the electron's magnetic moment is confirmed by experiments to within the instrumental precision of 1 part in a million, million (12 digits).

QUANTUM MECHANICS
UNDERLIES CHEMISTRY AND BIOLOGY

Quantum Mechanics provides an understanding of the atomic interactions that enable all the chemical and biological processes that sustain and enrich our lives. Let's see how this works by considering how two atoms interact to form a molecule. Recall from the last chapter that each element has a specific and unique set of electron energies, in accordance with quantum rules. Figure 17.1 shows the electron energy levels of sodium *Na* and chlorine *Cl*. Sodium has 11 electrons: 2 of these completely fill the lowest-energy orbit *n=1*; while 8 electrons completely fill the *n=2* orbit; and the last electron is alone in the *n=3* orbit. Chlorine has 17 electrons: again 10 of these fill the two lowest-energy orbits. This leaves 7 in the

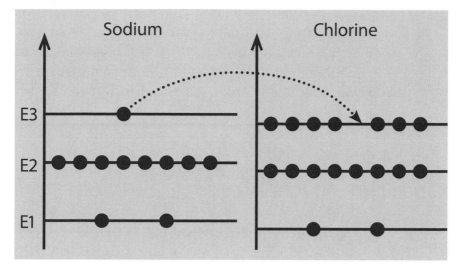

Figure 17.1. Black dots represent electrons occupying energy levels in sodium and chlorine atoms. The electron in the sodium's E3 level can reduce its energy by moving into the vacancy in chlorine's E3 level.

$n=3$ orbit, which can hold up to 8 electrons. Because chlorine has more protons than sodium (17 versus 11), the larger positive charge of the chlorine nucleus attracts electrons more forcefully than does sodium, drawing the chlorine electrons closer to the nucleus. Thus the energy of each orbit is somewhat lower in chlorine than in sodium; in particular, E3 in chlorine is lower than is E3 in sodium.

What this means is sodium's 11TH electron can be captured by the more forceful chlorine nucleus and moved into a lower-energy orbit. In nature, when something can reduce its energy, it almost always does. After the electron moves, the sodium atom has a net charge of +1 and the chlorine atom has a net charge of –1. The two atoms stick together because, as we know, opposite charges attract. This electron transfer creates sodium chloride *NaCl*, which is ordinary salt.

If the micro-world did not conform to the rules of Quantum Mechanics, every electron would collapse into the nucleus of its atom and there would be no molecules. Salt is a simple example, but even the most complex molecules exist because quantum energy levels determine how atoms combine to achieve the lowest possible electron energies.

In fact, at a fundamental level, all biological and chemical reactions occur because the electrons move to more favorable energy levels. This is why gasoline releases energy when it burns, why hemoglobin carries oxygen to our cells, and why DNA defines our genetic code. Quantum Mechanics provides the basis for understanding these reactions.

> If quantum mechanics did not reign in the micro-world, there would be no atoms, no planets, and no people.

QUANTUM MECHANICS AND ELECTRONICS

Quantum rules also enable modern electronics. Understanding the electron energy levels of silicon implanted with precise amounts of other elements allows us to design transistors that orchestrate the flow of electric current in integrated circuits. Integrated circuits are the building blocks of all digital electronics, including computers, cell phones, TVs, and MP3 players. Integrated circuits can contain a billion transistors in an area the size of a dime. The design of all modern electronics depends on Quantum Mechanics to provide a true understanding of the micro-world.

QUANTUM WAVE INTERFERENCE

We turn now to the intriguing Quantum Mechanics of wave interference. If a particle can reach a destination by two or more different paths, the particles-waves from each path combine to determine what happens at the destination. This leads to important interference effects in the micro-world that are unlike anything we experience in the macro-world. Wave interference is particularly significant when two waves have the same frequency and a fixed *phase shift*.

Phase shift is how much one wave form is shifted relative to another wave of the same frequency, as shown in the upper portion of figure 17.2.

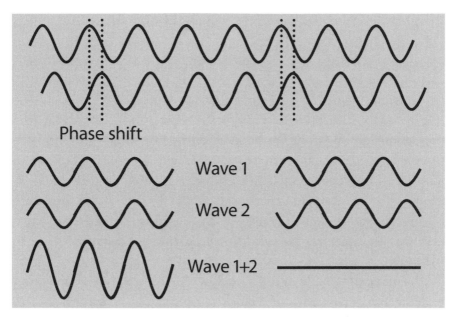

Figure 17.2. At the top, waves with the same frequency and an unchanging phase shift are called coherent—their crests are always separated by the same amount. Below and on the left, waves with zero phase shift interfere constructively. Below and on the right, waves with a half wavelength phase shift interfere destructively and cancel one another completely.

If two waves always crest at the same time, their phase shift is zero—they wave up and down together and they *interfere constructively*, as in the lower left side of that figure. If one wave always crests when the other troughs, their phase shift corresponds to half a wavelength and the waves *interfere destructively*; they totally cancel one another, as in the lower right side of that figure. Phase shifts can correspond to any fraction of a wavelength and the interference is then partial, being somewhere in between the two extremes of totally constructive and totally destructive.

Waves with the same frequency and a fixed phase shift are said to be *coherent*. Waves are *incoherent* if they have different frequencies or have varying phase shifts. Phase shifts will vary if any of the path lengths of the interfering waves vary. Only coherent waves with a one-half wavelength phase shift can totally cancel one another.

SCHROEDINGER'S INCOHERENT CAT

Let's now discuss a famous but much maligned creature: Schroedinger's cat—we'll call him Felix.

Austrian physicist Erwin Schroedinger received the 1933 Nobel Prize in Physics for developing a set of mathematical tools to calculate the predictions of Quantum Mechanics. Schroedinger and Einstein both made essential contributions to the development of Quantum Mechanics, and both were repelled by its uncertainty. In 1935, Schroedinger concocted a bizarre thought experiment intended to show that Quantum Mechanics leads to unreasonable conclusions. Rest assured that no real cats were injured by the politically incorrect thought experiment shown in figure 17.3.

The noble (not Nobel) Felix is enclosed in a sealed box with a radioactive source, a detector, and a bottle of poison gas. When the detector senses a radioactive decay, it releases the poison gas. Remember, this is a thought experiment that was never actually performed. Radioactive sources obey quantum rules; they decay at a precisely predictable rate, but each atom decays at a random, unpredictable moment. The probability of a decay increases steadily over time, but when a decay will actually occur is completely unpredictable.

Question: When the probability of detecting a decay reaches 50%, is Felix alive, dead, or both? *Both?* Who would possibly say both? Quantum Mechanics would.

Bohr would say Felix is in a quantum combination of both alive and dead states, with equal probabilities of each. Felix is neither alive nor dead, but is in a combined state until the box is opened and we observe him. At that instant, Felix's combined state "collapses" and he is then either alive or dead, but no longer both. Bohr said "Nothing exists until it is measured." In this context, observation is a form of measurement.

Einstein and Schroedinger thought that was crazy! Felix is either alive or dead with equal probabilities, but certainly he cannot be both. Also, it cannot make any difference whether or not we observe him. There is a reality independent of our observation. As Einstein said: "The Moon would not disappear if we stop looking at it."

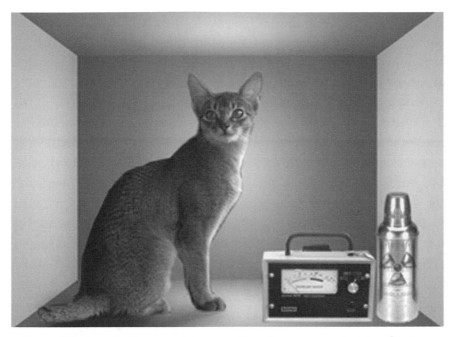

Figure 17.3. A sealed box contains: a radioactive source; a bottle of poison gas; a detector that releases the gas when the source decays; and Felix, Schroedinger's Cat. When the probability of a radioactive decay reaches 50%, is Felix: Alive? Dead? or Both? This was only a thought experiment. Felix lives on in quantum lore.

If Felix were a subatomic particle, Bohr's view would be correct. In the micro-world, subatomic particles really can exist simultaneously in combinations of very different states. This phenomenon, called *quantum superposition*, relies on coherence. But macro-world objects like Felix cannot persist in a quantum superposition of radically different states, because such states lose their coherence almost instantly.

Quantum superposition is the core technology in the ongoing development of quantum computers. This practical example will clarify Felix's true fate. In normal computers, numbers are stored in *bits*, each having a value of either 1 or 0. An *8-bit* device can store any *one* value from 0 to 255. In quantum computers, an *8-qubit* device can store *every* value from 0 to 255. A *32-qubit* device could store four billion values at the same time. Calculations could be done with all these values simultaneously.

This would expand computing power exponentially. While quantum computers are not yet practical, they have the potential to do what no current computer can: crack the secret codes of every bank, every army, and every nation. Enormous effort (and money) is being invested trying to make quantum superposition a practical reality.

The challenge impeding the development of quantum computers is maintaining coherence, even for a fraction of a second, even for devices with just a few atoms. So far, the state of the art is an eight-atom device, but its reliability is not yet adequate. Eight atoms is the very best that the world's most advanced technology can achieve.

Felix has 1,000,000,000,000,000,000,000,000,000 atoms (he's a Fat Cat). The multitude of interactions of particles in the air, the box, and Felix instantly disrupt any coherence between the alive-Felix and the dead-Felix states by altering the frequency of those states and introducing chaotic phase shifts. Quantum coherence cannot persist for a macroscopic object in a macro-world environment. Coherence is far too delicate to withstand the onslaught of the macro-world. Schroedinger's cat lives on in quantum lore—Felix has the last meow.

THE TWO-SLIT EXPERIMENT

The most famous example of interference is the two-slit experiment shown in figure 17.4. At the bottom of the figure, a source emits photons that hit a barrier with two slits that can be covered or uncovered by doors. Photons that pass through the slits are detected as they hit a flat surface at the top.

The figure shows three versions of the experiment with different combinations of door positions. With only the left door open, light diffracts in the left slit and produces a smooth distribution centered on the left side of the detector. Similarly there will be a smooth distribution on the right if only the right door is open. But with both doors open, light's wave properties create an interference pattern at the detector. The pattern consists of alternating bands of high intensity (white) and zero intensity (black) that are called *interference fringes*. Why?

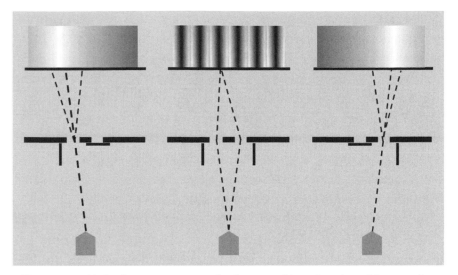

Figure 17.4. Light from sources at the bottom hits a barrier with two slits and doors. With only the left door open, light diffracts and illuminates a broad area on the left side of the detector at the top. Similarly, if only the right door is open. With both doors open (center image), light waves from the slits interfere, creating a fringe pattern

The source is placed equally distant from each slit, thus light waves from the source pass through each slit at the same time—with zero phase shift—they crest and trough at the same time. The interference at the detector is entirely due to differences in the path lengths from the slits to the various locations on the detector.

At the middle of the detector, the distance to the left slit *DL* is the same as the distance to the right slit *DR*. At the middle, the waves from each slit combine with zero phase shift. They crest and trough together and interfere constructively producing high-intensity wave energy.

Everywhere else on the detector *DL* and *DR* are not the same. Wherever *DL* and *DR* differ by an integral number of wavelengths, the waves have zero phase shift; they crest and trough together, resulting in high intensity. Where *DL* and *DR* differ by a half-integral number (½,³⁄₂, …) of wavelengths, one wave crests when the other troughs and they cancel one another completely, resulting in zero intensity. In between these two extremes, the phase shift is somewhere between zero and half a

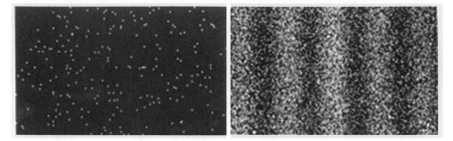

Figure 17.5. These images are the results of a two-slit experiment done with electrons instead of light. Left image, with 200 electrons, shows they are particles as each makes a pinpoint hit on the detector. Right image, with 140,000 electrons, shows they are also waves that produce an interference fringe pattern. Images courtesy of Akita Tonomura of Hitachi.

wavelength and the intensity will be somewhere between maximum and zero. Moving across the detector, the path length difference **DL–DR** gradually changes and the intensity oscillates, thereby producing the fringe pattern.

The locations on the detector with zero intensity are the most interesting. These are places that light would have reached if only the left door were open or would have reached if only the right door were open, but cannot reach with both doors open. Opening a second door prevents light from the first door from reaching the zero-intensity locations. Having a second way to get there prevents light from getting there at all!

Well, by now we know waves interfere and do other strange things. But astonishingly, so do particles, as shown in figure 17.5. These are real images, made not by light but by electrons. Notice the interference fringes on the right and the pinpoint spots on the left. Since this is a much more difficult process, the image quality is not as good with electrons as with light. Nonetheless, interference fringes are certainly evident in the image on the right that is a long-duration exposure. The left image is a brief exposure, with only 200 electrons. It confirms the particle nature of electrons as each electron hits a single pinpoint spot on the detector. This is particle-like behavior; waves don't hit only a single point. (Ever see an ocean wave crash onto the beach and hit only a single grain of sand?)

The images in figure 17.5 were taken with such a low-intensity source that only one electron at a time passed from the source to the detector. We must accept the startling conclusion that each electron-wave goes through both slits simultaneously, producing waves that interfere with one another, and nonetheless hit only a pinpoint spot on the detector! Only Quantum Mechanics has the audacity to predict such bizarre behavior by particles. Yet, this incredible image is proof that it really does happen, exactly as Quantum Mechanics predicts.

Feynman said that all the weirdness of Quantum Mechanics is encapsulated in the two-slit experiment.

VIRTUAL PARTICLES

Virtual particles are another remarkable quantum phenomenon. Recall Heisenberg's uncertainty equation: $dt \times dE > h/2\pi$, where h is Planck's constant, a number that sets the scale of Quantum Mechanics. This equation allows small, momentary deviations from perfect energy conservation. A small amount of energy dE can spontaneously appear from nothing if it disappears within a time dt, provided that $dt \times dE < h/2\pi$. This is because nature does not allow so small an energy deviation to be detected in so small a time. This is not a limitation of our instruments; it is due to the intrinsic uncertainties of energy and time due to the wave nature of particles in the micro-world.

Because h is very small, only an infinitesimal amount of energy can be "borrowed" for an infinitesimal amount of time. (You can't borrow a billion kWhs and hope to return it many years later in your will.)

It is possible for a particle and an antiparticle to spontaneously pop into existence from "absolutely nothing." In Quantum Mechanics, even the nothing of empty space is uncertain; nothing is truly "absolutely nothing." Particle-antiparticle pairs can exist for only a very brief time and then disappear together, leaving behind the original nothing, as shown in figure 17.6. These ghostly incarnations are called *virtual particles*. A virtual electron-antielectron pair can "exist" for about 10^{-21} seconds, not long enough to travel the width of an atom at the speed of light.

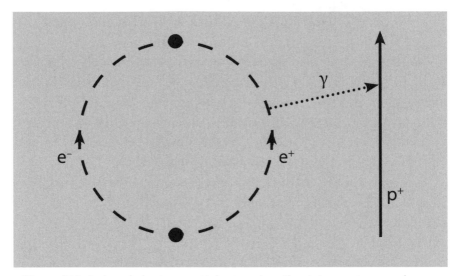

Figure 17.6. A virtual electron-antielectron (e⁻ e⁺) pair is spontaneously created in the black dot at the bottom and later recombines in the black dot at the top. During their brief existence, virtual particles can interact with real particles and change the real particles' behaviors. In this case, the antielectron emits a photon (γ) that is absorbed by a proton (p⁺).

Virtual particles may seem outlandish, but they really do exist and they cause very real effects that have been precisely confirmed. Virtual particles can interact with real particles and change the behavior of the real particles. For example, virtual particles increase the magnetic moment of all electrons by a factor of 1.001,159,652,181. This factor has been measured to a precision of ±1 in the twelfth and last decimal digit and confirms the virtual particle effect predicted by Quantum Mechanics to 1 part in a billion.

Any particle-antiparticle pair can be spontaneously created as virtual particles, although pairs with lower masses are more probable. As we discussed in chapter 7, a particle-antiparticle pair has zero net characteristics other than energy, all other characteristics cancel one another. These particle-antiparticle pairs can be created from pure energy and will annihilate to produce pure energy. In the case of virtual particles, they are created from energy "borrowed" from the Bank of Heisenberg on a short-term loan.

IDENTICAL PARTICLE EFFECTS

A very important aspect of Quantum Mechanics and elementary particles is the behavior of identical particles. When two particles are *exactly* identical, so identical that nothing in nature can distinguish them in any way, special quantum rules apply that dictate their behavior. There are two opposite behaviors in nature: the antisocial behavior of identical fermions and the gregarious behavior of identical bosons.

Fermions include all the primary constituents of matter: electrons, protons, and neutrons. They obey *Fermi-Dirac statistics* and are antisocial. No two identical fermions will ever share the same state; each demands its own turf.

Because each electron in an atom does demand its own state, they cannot all occupy the prime location—the lowest energy level, which is the $n=1$ orbit. (See figures 16.2 and 17.1.) When all the possibilities are properly counted, we find there is enough turf for two electrons in orbit $n=1$, eight electrons in orbit $n=2$, etc. These quantum rules are the basis for the structure of the Periodic Table and for the chemical properties of each element.

Bosons include all the carriers of forces, such as photons. They obey *Bose-Einstein statistics* and are gregarious. The more photons that exist in a quantum state, the more other photons "want" to join them (the probability of a photon joining an existing group of photons increases dramatically with the number already in the group).

Einstein explained theoretically how this property of bosons could be used to produce coherent beams of vast numbers of photons. In such a coherent beam, the photons all have the same frequency and are all moving in the same direction. Thirty years later, American physicists Theodore Maiman, Arthur Schalow, and Charles Townes used Einstein's ideas to produce the first lasers and related devices.

"Life is like a bicycle,

to keep your balance,

you have to keep moving."

— Albert Einstein

PART 2

Stars

Life of Stars

General Relativity

White Dwarfs

Neutron Stars

Black Holes

Space Telescopes

18

Twinkle, Twinkle Little Star...

Stars are the building blocks of the universe. They provide heat, light, and materials necessary for life. Without stars, the universe would be bitterly cold, dark, desolate, and lifeless. Stars make the universe beautiful and habitable.

Shortly after the **Big Bang**, when the universe came into existence, hydrogen and helium accounted for all but one atom in a million. There was just a dash of lithium, but not a single atom of carbon or oxygen in the entire universe. Stars created the atoms life requires.

In regions far from the warmth of stars, the temperature of our universe is now –455 °F, which is 2.7 degrees on the Kelvin scale (written 2.7 K). Water freezes at 273 K, thus the temperature of the universe is 100 times colder than freezing and just slightly above the lowest possible temperature, *absolute zero*, 0 K. At absolute zero, there is no heat energy and all atomic motion and all chemical reactions cease. Stars provide the warmth necessary for life.

As dismal as the average temperature of the universe is, its density is even more discouraging. The average distance between oxygen atoms is 60 feet and for carbon atoms it is 70 feet. At these densities, one would need to gather all the atoms within a volume 8 billion times larger than Earth in order to make a single human being. Stars bring atoms together in sufficient numbers for life to exist.

The life and death of stars is in turn determined by the interplay of gravity and pressure. Gravity, driven by a star's mass, strives to crush it, while pressure, created in a star's core, tries to blow it apart. The pressure derives from the properties of the elementary particles. Since each star is made of the same types of particles, the only variable factor is the star's mass.

> The saga of each star is determined by its mass.

Stars are balls of gas so immense that their self-gravity sustains the ultimate fire: nuclear fusion. Fusion creates the light that sustains life and makes stars twinkle in the sky. Fusion in stars also creates the atoms from which planets and all living creatures are made.

A STAR IS BORN

Stars are born when immense clouds of gas collapse. Since all types of matter attract one another by gravity, one might think a gas cloud would immediately collapse because all its particles would fall directly into its center. Actually, collapse is surprisingly rare. Even now, 14 billion years after the Big Bang, only 10% of the atoms in the universe are in stars. One reason that more matter has not collapsed is that the rapid expansion of the universe spreads matter throughout an ever-increasing volume. Another reason is that gas clouds are typically vast and of nearly uniform density. These clouds can remain relatively unchanged for hundreds of millions of years or more.

Gas clouds that are illuminated by stars are called *nebulae,* and are some of the most beautiful sights in the heavens. Color Plate 5 is an image of the Eagle Nebula (shown in green, in the upper left corner). The wing span of this eagle is 200 trillion miles, about a million times the diameter of Earth's orbit around the Sun. Our entire solar system is smaller than the smallest dot in this image.

Our galaxy, the Milky Way, contains about 10,000 immense gas clouds, each with an average mass of 100,000 times the mass of our Sun. From here on, I will denote the mass of our Sun by *Msun*. Most cosmic masses are measured in multiples of *Msun*.

Gas clouds are quite stable, but nothing in the universe lasts forever. Galaxies often cluster in large groups and sometimes pass very close to one another, or even collide. Since galaxies contain hundreds of billions of stars, the immense gravitational shock of a galactic close encounter often disrupts gas clouds in both galaxies. Supernovae, the explosions of massive stars, also create shock waves that disrupt gas clouds. Whatever the cause, once part of a cloud becomes substantially denser than average, its tranquil balance ends. Denser regions have stronger gravity and pull in surrounding matter. This further increases their mass and leads to gravitational collapse and the formation of a cluster of new stars.

Color Plate 6 shows two new star clusters. NGC 602 is in the Small Magellanic Cloud, a satellite galaxy of our Milky Way that is 200,000 light-years away. These new stars are less than 5 million years old. NGC 3603 is in our galaxy, 20,000 light-years away; its stars are less than 2 million years old. Both star clusters are virtually brand new in astronomical terms. Recall that a light-year is a unit of distance—the distance light travels in one year—about 6 trillion miles.

New stars are also forming in collapsing gas clouds in the Pillars of Creation, shown in Color Plate 7. The bright pink spots are called EGGs, short for Evaporating Gaseous Globules. (One of the challenging tasks in space science is coming up with clever TLAs, Three Letter Acronyms. NASA excels at creating TLAs.)

The stars in a cluster can have a wide variety of masses, but all formed at about the same time and all are about the same distance from Earth.

As gas falls into a collapsing region, it heats up and creates pressure that could eventually stop further collapse. This heat energy must be dissipated if the cloud is to continue collapsing. New stars form only if the gas can radiate away enough energy, and if the gravitational imbalance is sufficient. The presence of heavier elements, such as carbon and oxygen, greatly assists in radiating heat and facilitating star formation.

These heavier elements come from the explosive deaths of previous generations of stars.

As gas clouds collapse, they spin faster and faster, just as figure skaters do when they pull in their arms and legs. Collisions among the collapsing gas particles flatten the gas into a disk. Particles in oblique orbits cross the disk twice per orbit and collide with particles in the disk. These collisions gradually equalize particle velocities. Eventually, most particles fall into the disk. It's easier to go with the flow than to swim up stream. As the disk grows denser, its own self-gravity flattens it further.

The entire collapse process can take hundreds of thousands of years. Eventually, most of the gas collapses into a small central ball called a *proto-star.* It is small only compared to the original gas cloud. Collapse can be relentless and efficient. Nearly 99.9% of the mass of our solar system collapsed into the Sun; the remaining gas formed planets, moons, asteroids, comets, and assorted debris.

Proto-stars with masses less than 8% of *Msun* don't have enough gravity to sustain fusion and they never become true stars; these are called *brown dwarfs.* In a proto-star with a mass greater than 8% of *Msun,* gravity squeezes its gas until its central temperature is high enough to initiate nuclear fusion. Stars just barely massive enough to sustain nuclear fusion are very dim and are called *red dwarfs.* Our Sun is more massive than an average star; it is a *yellow dwarf* and its core temperature is now 30 million °F. Today, the heaviest stars have masses of up to 150 *Msun* and are called *blue-white supergiants.* When the universe was much younger and much denser, some stars may have had masses of up to 1000 *Msun.* But such stars are long since gone, as we will discuss shortly. Color Plate 8 shows a comparison of star sizes, masses, and colors. As with most celestial entities, there are far more small stars than large stars. You may have noticed from that image that blue-white supergiants are much smaller than *red giants.* This is because the names *dwarf* and *giant* are not assigned based on size. Only a few stars are so close that astronomers can directly measure their actual size. Rather, astronomers call dim stars *dwarfs* and bright stars *giants,* without regard to what their physical sizes might actually be.

A LONG, EXQUISITELY BALANCED LIFE

When nuclear fusion ignites in its core, a proto-star becomes a true star. In addition to enormous light output, stars also produce formidable *stellar winds*, streams of high-energy particles. Powerful stellar winds from new stars blow away any surrounding gas—gas that has not already condensed into a large body such as a planet. An example of this is seen in the Helix Nebula shown in Color Plate 9. Here, an energetic new star has cleared its surroundings. The left side is a magnified image of the interior edge of the Helix's gas envelope. Loose gas has been blown away leaving only material that is sufficiently collapsed to be dense enough to withstand this intense erosion.

One million tons of particles are emitted each second in our Sun's solar wind. This is in addition to the 4 million tons of mass it converts into heat and light every second. Thus our Sun is on a weight-loss program, losing 5 million tons per second.

The stellar winds of new stars effectively prevent an entire gas cloud from collapsing into just a few stars. They prevent stars from becoming extraordinarily heavy and, therefore, short-lived and unsuitable for life. Nature produces a large number of moderately sized, long-lived stars around which life can develop and saves the remaining gas for later generations of stars.

Some say diamonds are forever; what about stars?

Stars are mortal and they too pass with time. But it can be a very long time indeed, from millions of years to millions of millions of years.

But why are stars long-lived at all? Why doesn't collapsing gas continue to collapse down to "nothing"? If you drop a rock in the ocean, it doesn't stop until it hits bottom. Why does the gas stop falling? It stops because of nuclear fusion in the star's core. Without fusion, our Sun would have burned out in a few thousand years and none of us would be here.

Nuclear fusion is nature's ultimate fire. As discussed in chapter 10, fusion depends on the properties of atomic nuclei and can produce

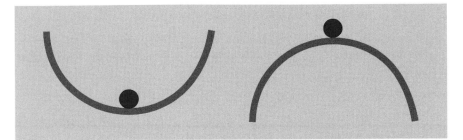

Figure 18.1. The left ball is in a stable equilibrium because any small change of position results in a restoring force (gravity) that pushes it back to the center. The right ball is in an unstable equilibrium because any small change of position leads to substantial and irreversible change.

millions of times more energy than a conventional fire. Fusion changes everything. When a star's core temperature rises to tens of millions of degrees, hydrogen in the core begins to fuse. This releases a tremendous amount of heat, leading to a correspondingly tremendous pressure that stops the star's further collapse.

Nuclear fusion gives stars a long and stable life, a long-lasting equilibrium, and a balance between gravity and pressure, as illustrated in Color Plate 10. Stable equilibriums can only exist where there are forces that restore balance when small changes occur, like the situation on the left side of figure 18.1. Should anything move the ball off center, gravity pushes it back down to the bottom, making this system stable. Should a star expand slightly, for any reason, its temperature, fusion rate, and pressure would greatly decrease. With less pressure to resist it, gravity would squeeze the star back down. Should a star contract slightly, its temperature, fusion rate, and pressure would greatly increase; the higher pressure would overpower gravity and re-expand the star.

This delicate balance can keep a star burning at a nearly constant rate for billions of years. Our Sun is a huge inferno, a million times Earth's size, with an extremely precise thermostat on which our lives depend. If the Sun's energy output—its *luminosity*—doubled, Earth would receive as much solar energy as Venus does now, with its average temperature of 800 °F. If the Sun's luminosity were halved, Earth might have the climate that Mars does now, with an average temperature of –80 °F.

Life requires much more favorable temperatures. The *habitable zone* of our solar system is generally considered to be the region where water can be liquid. While it is not absolutely impossible for some form of life to exist elsewhere, the sweet spot is right here. The habitable zone is a narrow band centered on Earth's orbit. For Earth to remain habitable, the Sun's luminosity cannot vary by more than ±10%, and perhaps much less. Multicellular life evolved on Earth in the last 600 million years, during which the Sun's luminosity varied only ±3%. Compare that variation to the full range of stellar luminosities: the brightest star is a billion times brighter than the dimmest star. Our Sun is near the middle of this range, with a thermostat that holds its luminosity stable to better than 1 part in a million of this range. If your home heating system were that precise, the temperature in your house would never vary by more than 1/10,000TH of a degree. We are in a *Goldilocks Zone*, where the temperature is stable, not too hot and not too cold, but just right.

Clearly, we've got it good. But how long will the good times last?

Stars are remarkably stable while hydrogen fusion continues in their cores. Our Sun's hydrogen fusion stage will last 10 billion years; we're now at halftime. During its 10 billion year life, a star with our Sun's mass converts 10% of its original hydrogen into helium. Being heavier, helium sinks to the center and pushes out the hydrogen. Eventually, helium starves the core of hydrogen fuel and ends the hydrogen fusion stage. More massive stars burn their hydrogen much more rapidly and die much sooner; less massive stars are very dim and live much longer. Stars 25 times more massive than our Sun burn hydrogen 36,000 times faster and run out of hydrogen in only 7 million years. Even though they start with so much more fuel, their lives are 1500 times shorter.

IS THERE LIFE AFTER HYDROGEN FUSION?

Regardless of its mass, every star ultimately runs out of hydrogen in its core. Its pressure wanes and gravity finally wins the epic battle with pressure. Gravity compresses the core and pushes its temperature ever higher. For stars with masses of at least ½ *Msun*, core temperatures eventually reach 300 million °F and the next stage begins: helium fusion. Three helium nuclei fuse to produce carbon. As the core reaches this enormous temperature, the resulting pressure pushes the star's outer layers farther out, dramatically expanding the star, and turning it into a red giant. Look again at Color Plate 8. When our Sun becomes a red giant, it will engulf Mercury and Venus, and may extend almost to Earth's orbit. We need to be out of here well before then.

When the supply of helium in the star's core is depleted, it may contract further and become even hotter. At higher core temperatures, additional fusion stages produce nitrogen, oxygen, and other elements. More massive stars reach higher core temperatures, progress through more fusion stages, and produce ever heavier elements. The most massive stars reach core temperatures of 7 billion °F, hot enough to produce iron. Each stage of nuclear fusion has a drastically shorter duration than the prior stage. The iron fusion stage lasts only one Earth-day—less than one-billionth as long as the hydrogen fusion stage.

Hence, a star's life is only slightly longer than the duration of its hydrogen fusion stage.

STARS DIE IN A BLAZE OF GLORY

Eventually, when all available nuclear fuels are consumed, a star's life ends. As fusion's heat and pressure wane, gravity relentlessly compresses the core. As the core contracts, gravity becomes even stronger because this force increases rapidly as distances shrink. Eventually, the core implodes catastrophically, releasing almost as much energy in one brilliant flash as the star may have produced in its entire life. The energy released by the implosion of the core blasts the star's outer layers out into space, along with the new atoms it created.

The most massive stars die in the most spectacular explosions called *supernovae*. One of the most famous supernovae is SN1987a. (Supernova are named "SN" followed by the year of their occurrence, followed by one or more letters; "a" is the first supernova of each year.) SN1987a occurred in the Large Magellanic Cloud, a satellite galaxy of our own Milky Way.

SN1987a exploded one million, million, million miles from Earth. But as far as that is, it was the nearest supernova since the invention of the telescope, so astronomers went wild. The first person to see SN1987a was an astronomer walking to work who happened to look up at the night sky. The supernova was bright enough to catch his eye even from that great distance and even without a telescope or binoculars.

For several weeks, some supernovae can be brighter than an entire galaxy. The amount of energy they release is truly incomprehensible. A supernova can release 10 million times more energy than the detonation of a stack of dynamite as large as our Sun.

RISING FROM THE ASHES

The cosmos is an efficient recycler. The spectacular death of a star sets the stage for new beginnings. Shock waves from stellar explosions often initiate the collapse of neighboring gas clouds and the formation of new stars. Vital elements created in stars are dispersed into the galaxy by the stars' violent deaths. The collapsed cores that remain are the most exotic objects in the universe: *white dwarfs, neutron stars,* and *black holes*, all of which are discussed in coming chapters.

Before their demise, massive stars can create all the elements with masses up through iron by the process of nuclear fusion. Because iron has the most tightly bound atomic nucleus, no energy can be released by fusing it into heavier elements. Once a star has finished producing iron, it can no longer generate the energy and pressure needed to prevent gravitational collapse, and it dies. Heavier elements, including silver, gold, lead, and uranium, are produced during the instant of the star's explosive death. When a stellar core implodes, it collapses to something dramatically smaller, releasing an immense amount of gravitational potential energy and increasing the star's luminosity up to 100-billion-fold. Some of that energy is absorbed in converting iron into heavier elements. Since these elements are less tightly bound than iron, fusing iron to create them absorbs energy from the blast.

Atoms heavier than helium, which are required for any form of life, are made only in stars and are dispersed by their explosive deaths. Interstellar gases enriched with these atoms later collapse to form new stars and new planets. This cycle is repeated over and over. Each generation of stars increases the amount of carbon and oxygen in the galaxy. Judging from the concentration of various elements in our solar system, the Sun may be a third-generation star: the material of our solar system may have come from the death of an earlier star that was itself made of material from the death of an even earlier star.

WE ARE 90% STAR DUST

All the atoms in Earth and in our bodies, other than hydrogen, were made in the centers of massive, extremely hot stars that exploded at least 5 billion years ago. The other 10% of our body weight, hydrogen nuclei, was produced in the first millionth of a second after the **Big Bang**; these atoms are almost 14 billion years old. But don't feel like Methuselah. Regardless of their age, atoms are always in pristine, brand-new condition.

Atoms are too small to have a clock—they never age, they never wear out, and they never run down. Atoms are timeless.

19

Newton and Einstein on Gravity

Gravity appears to be the simplest of nature's forces, perhaps because it is the most familiar. It may also seem incredibly weak compared with other forces because it is 10^{37} times weaker than the strong force. (For an explanation of what 10^{37} is, see note [1] at the end of chapter 5.) Yet gravity is in many ways the most complex and the most powerful force of nature. It controls the fate of the universe and everything in it.

Galileo Galilei was the first to study gravity scientifically. He discovered that, ignoring air resistance, all bodies fall at the same rate—a simple but powerful law. Galileo also discovered the moons of Jupiter, the first objects ever seen that unquestionably orbit something other than Earth. Rather than burn at the stake for heresy, Galileo reluctantly recanted some of his discoveries.

GRAVITY ACCORDING TO NEWTON

Sir Isaac Newton was the next to advance our understanding of gravity, publishing in 1687 his *Principia Mathematica*, thought by many to be the most influential book in science. Newton was prolific as a physicist, mathematician, alchemist, and theologian. His other scientific discoveries include his laws of motion, the reflecting telescope, and the theory of

Figure 19.1. Newton understood why the Moon orbits Earth. At each moment, the Moon moves forward in the direction of its current velocity (dotted arrow) and it also falls toward Earth (solid arrow). The sum of these two motions produces its orbital rotation around Earth.

color. However, Newton is said to have spent more time on his unorthodox biblical interpretations than on science. As England's Master of the Mint for 28 years, he reformed the currency, routed out counterfeiters, and greatly increased England's wealth and fiscal stability. For this, and not for his outstanding contributions to science, Queen Ann knighted him in 1705. A postmortem found very high levels of mercury in his body, no doubt due to his extensive work in alchemy. This could explain his "erratic" behavior in later life.

Newton and Gottfried Wilhelm Leibniz independently invented an entirely new branch of mathematics: *calculus*. This enabled Newton to formulate physical laws by relating forces to small changes, and relating sequences of small changes to global motion. This was the dawn of a new age of analytical science. Newton showed there is one law for all things gravitational: a universal law of gravity that works for apples, the Moon, and everything else. Never before did science have such sweeping reach.

Let's explore how Newton explained our Moon's orbit, as illustrated in figure 19.1. The dotted arrow indicates the Moon's velocity at a particular moment. An object's velocity is its speed and direction of motion, such as "60 mph due north." If it maintains that velocity for 1 hour, it will move 60 miles north. Maintaining that velocity for 1 minute moves it 1 mile north.

A stable orbit requires a balance of gravity and velocity. Without gravity, the Moon would forever move in a straight line in the direction of its velocity. It would leave Earth and never look back. But because of gravity, at every moment the Moon also falls toward the Earth, just as apples fall from trees, as indicated by the short solid arrow in the figure. If the Moon ever had zero velocity, it would begin falling straight toward Earth; its impact would vaporize our oceans and melt the planet's surface. Fortunately, what actually happens is the sum of both arrows; the Moon falls toward Earth and at the same time also moves forward with its current velocity. Each minute, the Moon moves forward 38 miles and drops 16 feet toward Earth. A 16-foot drop seems tiny compared with 38 miles, but it's exactly the right amount to turn the Moon from its straight line path into its orbit around Earth. By staring at apples, Newton understood the motion of the Moon. (Ain't physics cool.)

Newton said that gravity is caused by mass and that only mass responds to gravity. Since we now know light is made of photons that have zero mass, Newton's laws say light is not affected by gravity. He also said changes in the positions of massive objects are felt instantaneously throughout the universe. For example, if the Sun vanished, Earth would stop feeling its pull immediately. For that to be true, whatever "causes" gravity must travel with infinite velocity, which doesn't seem reasonable. And by the way, what is it that actually "causes" gravity? What is the mechanism by which the Sun reaches out and pulls on Earth across 93 million miles of empty space? No one knew, not even Newton. This mysterious, unseen mechanism is called an ***action-at-a-distance***. Not understanding gravity's mechanism bothered physicists, including Newton. While Newton had explained remarkably well what gravity ***did***, he had not explained ***how*** gravity worked. Nevertheless, Newton's theory was a tremendous achievement and remained the flagship of science for more than 200 years.

GRAVITY ACCORDING TO EINSTEIN

Einstein changed everything. He said gravity is not a force after all, but is the result of the geometry of our universe being curved. If the geometry of the universe were Euclidean, rather than curved, there would be no gravity and all objects would move in straight lines. Earth orbits the Sun, Einstein said, because the Sun curves the geometry of the solar system and Earth follows the straightest possible path in that curved geometry, as shown in figure 19.2.

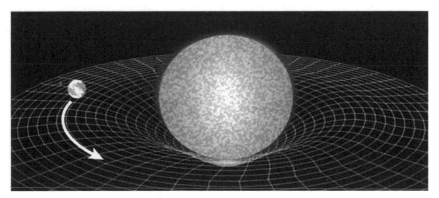

Figure 19.2. Einstein said that gravity is not a force but is the result of the curvature of spacetime. The Sun curves the geometry of our solar system, somewhat like a bowling ball would deform a bed sheet. The Earth follows the straightest possible path through this curved geometry—an orbit around the Sun.

Figure 19.2 shows only two dimensions of our universe's four-dimensional spacetime. No one I know can draw, or even imagine, all four dimensions at once; we can only draw part of the geometry and hope that conveys the key ideas. In this figure, our universe is represented by only the deformed two-dimensional surface with crisscrossed white lines. To make them easier to see, the Sun and Earth are shown as balls laying on that surface, but they should really be drawn as flat disks entirely within the two-dimensional surface of our universe. Anything outside this two-dimensional surface is outside our universe.

Perhaps our universe bends in a fifth dimension, corresponding to the vertical direction of this figure. That larger space, with five or more dimensions, is called *hyperspace.*

Euclidean geometry is what we all learned in high school: parallel lines never cross, and all that. All this is true on a *flat* surface, like a sheet of paper, but it is not true on a curved surface, like the surface of a sphere. For example, on Earth's surface, the meridians are parallel at the equator, but intersect at both poles. I will say more about curved geometries in the next chapter.

Clearly Einstein's concept is very different from Newton's, as shown in figure 19.3. Einstein said that all forms of energy, not just mass, cause gravity and that gravity affects all forms of energy, not just mass. Since Einstein previously said mass and energy were equivalent, we should have seen that coming. In Einstein's gravity, there is no action-at-a-distance; the Sun curves geometry where it is, and Earth responds to the geometry where it is. Geometry is the mechanism that links the Sun and Earth; curving the geometry in one location affects the geometry everywhere (like pulling on one end of a bed sheet).

Einstein said that changes in gravity are really changes in geometry. He called these changes *gravity waves* that ripple through space and time like ripples on the surface of a pond, except that gravity waves travel at the speed of light. If the Sun were to vanish, we would continue to see its light and feel its gravity for another 500 seconds because that's how long it takes light and gravity to travel 93 million miles. After 500 seconds, both the Sun's light and its gravity would vanish together.

Einstein also changed our entire understanding of space and time. Newton thought there could never be any disagreements about the length of a mile or the duration of a second because he believed space and time were absolute and fixed—the same for every observer, everywhere and always. He also believed space and time were two completely unrelated entities. Einstein showed that space and time are relative—different observers measure different values for distance and time, and that space and time are intimately united as one entity: *spacetime.* They are really two sides of the same coin. When one changes reference frames, some of what was space can become time and some of what was time can

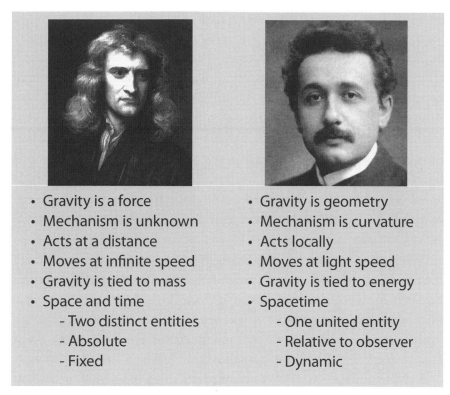

- Gravity is a force
- Mechanism is unknown
- Acts at a distance
- Moves at infinite speed
- Gravity is tied to mass
- Space and time
  - Two distinct entities
  - Absolute
  - Fixed

- Gravity is geometry
- Mechanism is curvature
- Acts locally
- Moves at light speed
- Gravity is tied to energy
- Spacetime
  - One united entity
  - Relative to observer
  - Dynamic

Figure 19.3. The two greatest physicists had very different theories of gravity, as shown by this comparison. Left: Sir Isaac Newton (1642–1727)

become space. In Einstein's Theory of General Relativity, spacetime is dynamic—the geometry of spacetime is constantly changing as mass and energy move.

Newton viewed space and time as a fixed stage on which the drama of the universe is played out. Einstein viewed spacetime as a dynamic stage that is part of the drama and that controls the motions of the actors.

Einstein explained why, ignoring air resistance, all freely-moving objects fall at the same rate, as Galileo had discovered. This is because all freely-moving objects travel through the same curved, four-dimensional geometry, along the same curved paths—paths that have nothing to do with their masses. In fact, light, which has zero mass, also moves along these same curved paths. Therefore, Einstein said, light bends as it passes a massive body like the Sun.

GRAVITY BENDS STARLIGHT

According to Einstein, starlight passing just above the Sun's surface bends by 1/2000TH of a degree. Light passing twice as far from the Sun's center bends half as much. Consider the situation shown in figure 19.4. Normally, the Sun is not in our line of sight to star A; then, light from that star is not bent by the Sun and we observe the star in its actual location. However, when the Sun moves into our line of sight, light from star A bends as it passes very close to the Sun. As we look back along the light ray that reaches us, the light appears to come from location A* instead of location A. Therefore, we see star A appearing to be closer to star B than it actually is.

When Einstein published General Relativity in 1915, the small bending angle he predicted could be measured (barely), but the challenge was that this measurement could only be done during a total solar eclipse. Stars near the Sun are visible only when the Moon blocks the Sun.

Total solar eclipses are infrequent, brief, and localized. They occur about once every 18 months, often in inaccessible places. Totality lasts only a few minutes, and covers only a tiny sliver of Earth's surface. At any one location, a total solar eclipse occurs only once every 370 years. Astronomers have little chance to overcome error or bad luck—if the sky is cloudy, the opportunity to see stars near the Sun is lost.

The solar eclipse of 1919 was centered in the South Atlantic. British astronomer Sir Arthur Eddington, shown in figure 19.5, made an extraordinary effort to get the most from this less than ideal opportunity to test Einstein's revolutionary theory. Eddington, a Quaker and a conscientious objector, refused military conscription, which wasn't well-accepted during the war. Only with the aid of powerful friends was he able to avoid prison and continue his research. Eddington led expeditions to both sides of the South Atlantic, to Brazil and to the island of Principe near Africa. During the eclipse, his teams feverishly took picture after picture, through partially cloudy skies, hoping at least one would succeed.

Standing under an immense portrait of Sir Isaac Newton, Eddington formally announced his findings at a meeting of Britain's Royal Society. He declared that Einstein's prediction of the bending of starlight was

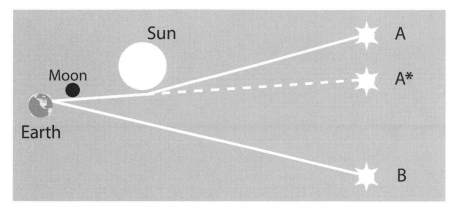

Figure 19.4. During a total solar eclipse, light from star A bends as it passes the Sun. We see that light as coming from location A*—star A appears to be closer than normal to star B, a star whose light is bent much less.

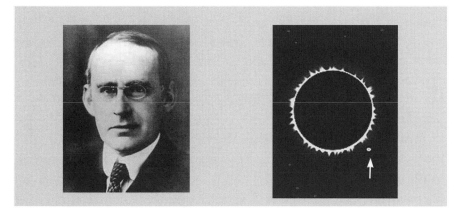

Figure 19.5. Sir Arthur Eddington (1882–1944) and an image from the 1919 solar eclipse. The indicated star's position confirms General Relativity.

correct and that Newton's theory of gravity must give way to Einstein's Theory of General Relativity. Again, a better understanding of gravity ushered in a new age of science.

It is a tribute to international scientific cooperation that near the end of the catastrophe that was World War I, a British scientist devoted so much effort to test a radical revision of Newton's theory of gravity that was proposed by a German scientist.

20

Einstein's Theory of General Relativity

Einstein's Theory of Special Relativity, published in 1905, applies only when everything moves with a constant velocity—when there are no forces and no accelerations. In particular, Special Relativity assumes that the effects of gravity are negligible. While this theory revolutionized our understanding of many important natural phenomena, there was clearly a need to extend the theory to include forces and, most importantly, to incorporate gravity.

Einstein succeeded in extending Special Relativity, publishing his Theory of General Relativity in 1915. General Relativity applies *generally*—to all states of motion, with or without forces, and incorporates gravity. We believe General Relativity is valid in all but two circumstances. It fails only where all our current physical theories fail: (1) at the center of black holes, and (2) at the instant the universe began. In these two circumstances, the mathematics on which all our physical theories rely is not applicable. There, space and time appear not to be continuous and smooth, as calculus requires. We will dig into this further when we discuss black holes (chapter 25) and cosmology (chapter 39). If we stay away from these two extremes for now, Einstein's Theory of General Relativity is valid for any and all circumstances.

Because General Relativity encompasses any state of motion, including acceleration, an inertial frame is no longer required. The Principle of Relativity can now be restated (we'll discuss the meaning of ***properly written*** later):

> ### The Principle of Relativity
>
> The Laws of Nature, when properly written,
> are the same for all observers, regardless of their motion.

A critical step in developing General Relativity occurred in 1907, while Einstein was still a patent clerk. He realized that someone in free fall would not feel gravity. Consider a physicist in an elevator at the 100TH floor, and imagine a cable suddenly breaks dropping the elevator. While this poor soul haplessly falls toward a sudden stop, he sees no evidence of gravity. Naturally, our physicist is eager to perform numerous (quick) experiments to verify this. He grabs his keys, extends his arm, and releases them; the keys seem to hang motionless in space. Then he places his keys on the ceiling, and watches as they stay there. And then he… (oops).

Einstein called this realization his "happiest moment" (not the "oops" part). To Einstein, this insight meant free fall is the simplest and most natural state of motion. (Being free of gravity does not, however, free physicists from the struggle for tenure.) It is the floor that makes us deal with gravity—by preventing us from falling, the floor makes physics seem more complicated than it really is. Einstein developed this realization into the fundamental principle of General Relativity, the Equivalence Principle:

> ### The Equivalence Principle
>
> Locally, gravity is equivalent to a uniform acceleration.

Here, "locally" means a small region of space.

Earth's gravity always pulls things down toward its center. But "down" is in different directions in London, Vancouver, and Sydney.

Gravity also varies with the distance to Earth's center; it's slightly less atop Mt. Everest than on the Dead Sea. But within a small enough region, we can regard gravity as being uniform—as pulling in the same direction with the same strength everywhere. Within such a region, "locally", the Equivalence Principle states that gravity could be replaced by a uniform acceleration without altering the motions of any objects, the outcome of any interactions, or the results of any measurements.

The Equivalence Principle made developing General Relativity easier mathematically. With this breakthrough, it took Einstein "only" another eight years to complete his theory.

To better understand this important idea, let's consider another example: firing a cannon in four situations, as shown in figure 20.1. In each of the four parts of the figure, the dotted line indicates the height of the cannon barrel at the moment it fires, and the cannonball's position, as seen from the cannon's viewpoint, is shown at 1, 2, 3, and 4 seconds after firing. Part (a) shows what would happen if there were no gravity: the cannonball doesn't fall. It travels in a straight horizontal line, moving the same distance each second. This is about as simple as any motion can be. Part (b) shows what happens with gravity: the cannonball moves horizontally as before but also falls ever more rapidly due to gravity. The path is more complex: it's a parabola. In part (c), the cannon and the ball are both allowed to fall freely in normal gravity. As Galileo discovered 400 years ago, the cannon and the ball fall at the same rate. From the cannon's viewpoint, the ball moves forward in a straight line and is always at the same height as the cannon. The motion appears exactly as in part (a) where there is no gravity. As Einstein said, in free fall there appears to be no gravity. Finally, in part (d) we again eliminate gravity, put a rocket under the cannon, and accelerate it upward at 32 feet per second per second (the same acceleration as Earth's gravity). The cannon rises faster and faster due to the rocket, while the ball moves in a straight line since, without gravity, there is no force acting on it. But from the cannon's viewpoint, the ball is falling faster and faster (velocity is relative) and appears to follow exactly the same parabolic path as in part (b). As Einstein said, gravity can be replaced by a constant acceleration.

This is the Equivalence Principle.

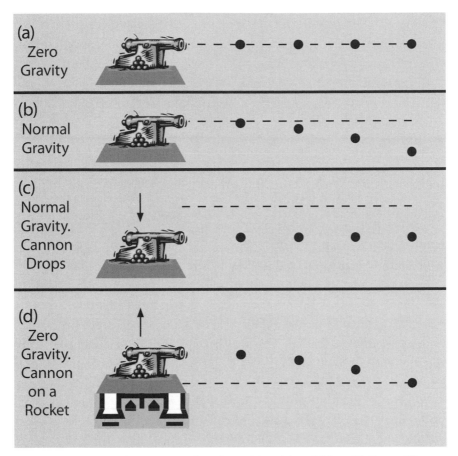

Figure 20.1. Equivalence Principle relates (a) to (c) and (b) to (d). Dotted lines show height of cannon barrel when fired. Position of cannonball is shown at 1, 2, 3, and 4 seconds after firing, as seen from the cannon.

SPACETIME CURVATURE

Perhaps the greatest challenge in understanding General Relativity is grasping the concept of *spacetime curvature*. Einstein said that what we call gravity is instead the effect of curved geometry. What exactly is curved geometry?

Euclidean geometry has zero curvature. Topologists use the word *flat* to describe any space with zero curvature, even spaces with three or more dimensions. In any flat geometry: parallel lines never cross;

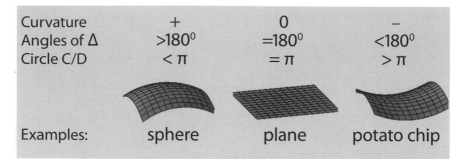

Curvature	+	0	−
Angles of Δ	>180°	=180°	<180°
Circle C/D	< π	= π	> π
Examples:	sphere	plane	potato chip

Figure 20.2. Above is a comparison of flat and curved two-dimensional surfaces. On a flat surface, such as a plane, the curvature is zero, the sum of the interior angles of any triangle is 180 degrees, and the circumference C divided by the diameter D of any circle equals π. None of this is true on a surface with positive curvature, such as a sphere, or a surface with negative curvature—a hyperbolic surface—such as a potato chip.

the sum of the interior angles of any triangle is 180 degrees; and the circumference of any circle is π times the diameter—all the good stuff we learned in high school. All this is true on a flat surface, like a sheet of paper, but *none* of this is true on a curved surface. See figure 20.2.

On surfaces with positive curvature, like a sphere, all triangles contain more than 180 degrees and the circumferences of all circles are less than π times their diameters. On surfaces with negative curvature, called *hyperbolic,* all triangles contain less than 180 degrees and the circumferences of all circles are more than π times their diameters. As figure 20.2 shows, a hyperbolic surface looks like a potato chip, curving up in one direction and curving down in the perpendicular direction. Surprisingly, the surface of a cylinder is *flat* because triangles and circles on a cylinder have the same properties as on a plane. Rolling a sheet of paper into a cylinder doesn't stretch the paper; no curvature is introduced going from a flat sheet to a cylindrical surface.

Figure 20.3 shows a triangle drawn on Earth's surface that has three 90-degree angles. Starting at the North Pole, two meridians are selected that are 90 degrees apart. Each meridian intersects the equator at 90 degrees, for a total of 270 degrees. As 270 is more than 180, Earth cannot be flat, it must have positive curvature. Every triangle drawn on Earth's surface has more than 180 degrees. In fact, a 16-year-old mathematician told me that on a sphere the internal angles of a triangle can add up to as much as 900 degrees. (Can you guess which triangle that is? [1])

Any circle on Earth's surface has a circumference less than π times its diameter. Consider the equator, a circle with a circumference of 24,902 miles, as shown in figure 20.4. In our usual flat, three dimensions, the diameter of the equator runs through Earth's center and is 7926 miles long. The circumference divided by the diameter is π; Euclid would be happy. But viewed in the curved two-dimensional space of Earth's surface, the diameter of the equator runs over the North Pole. The shortest diameter that remains entirely within the two-dimensional surface is 12,451 miles long. In this curved two-dimensional space, the circumference of the equator is only twice its diameter.

The curvature of any space, even a very irregular space, can be precisely determined by measuring triangles or circles within that space. It is not necessary to be outside the space to determine its curvature. Good thing, since we want to know the curvature of our universe and we cannot go outside it and look back.

Figure 20.3. Image shows a triangle drawn on Earth's surface that has three right angles, for a total of 270 degrees; therefore Earth's surface is not flat, it has positive curvature.

Figure 20.4. On the two-dimensional surface of the Earth, the diameter of the equator runs across the North Pole, and is half of the equator's circumference. In three dimensions, the diameter runs through Earth's center, a shorter route.

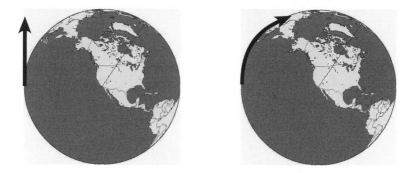

Figure 20.5. Left: From any point on Earth's surface, anything moving on a straight line goes off the planet. Right: The straightest possible lines within a curved surface are geodesics; on Earth they are called great circle routes.

In a curved space, there are no straight lines. Anyone moving on a straight line on Earth goes off the planet and out of its surface, as figure 20.5 shows. The straightest possible line within a curved space is called a *geodesic*, which is the shortest distance between two points in a curved space. On Earth's surface, geodesics are called great circle routes.

So what about gravity? Einstein said gravity is not a force. We exist in a four-dimensional spacetime that is curved by mass and all other forms of energy. Free objects move through spacetime along the straightest possible paths: geodesics. Here "free" means not subject to any of the other forces of nature: strong, weak, or electromagnetic. Figure 19.2 illustrates the spacetime curvature caused by the Sun and how Earth moves in this curved space, something like a ball rolling around the inside of a bowl. There are an unlimited number of great circle routes passing through each point on Earth, one heading in every possible direction. Similarly, through every *event* in four-dimensional spacetime, there is a geodesic for every possible velocity an object might have.

As figure 20.6 shows, having the curvature of space control the motion of objects is not unfamiliar. Anyone who enters the top of the slide is forced to follow its curving path to the bottom—the only way to go down is to also turn around. While it's much harder to visualize, similar logic applies in four-dimensional curved spacetime. For a ball released from the Leaning Tower of Pisa, the only way to go forward in time is to drop

Figure 20.6. Curved spaces control the motion of objects within them. For an object at the top of the slide, the only way to go down is to turn around.

at an ever faster rate. For the Earth to go forward in time, it must orbit the Sun.

We will now explore General Relativity in more depth. In order to make the discussion simpler, I will follow a standard convention and adopt *natural units*. Physical quantities are measured according to some chosen set of units. For example, speed can be measured in miles per hour, meters per second, or even fathoms per fortnight. We might as well choose units that make life easier. The best choice is natural units in which the speed of light c and Newton's constant G are both equal to 1. This way c and G disappear from the equations and make them simpler. You might ask, since the speed of light is enormous, how can we make $c=1$? It's easy: measure distances in light-years and measure time in years. As speed=distance/time, c=(1 light-year)/(1 year)=1. By choosing the right units of mass, we can also make $G=1$. This may seem odd, but trust me, it is simpler and that helps. Using natural units comes just in time as we need G for something else.

A noted mathematician once said: "Life is short, but the alphabet is even shorter." Science long ago used up all the letters of the alphabet, and all the Greek letters as well—we need to recycle.

EINSTEIN'S FIELD EQUATIONS

Einstein's Theory of General Relativity is embodied in his Field Equations:

$$G = 8\pi T$$

If you want to learn just one equation from this book, this is it. Here, **G** is the **Einstein tensor** and represents the curvature of spacetime. **T** is the **energy tensor** and represents the entirety of all forms of energy, mass, and pressure. As described by the renowned American physicist John Archibald Wheeler, Einstein's Field Equations state: "Mass and energy tell space and time how to curve, and space and time tell mass and energy how to move."

Einstein demonstrated that nature cannot be fully understood within the context of time and three-dimensional space. We must view nature in the context of curved, four-dimensional spacetime. *Tensors* are mathematical entities that conform to the rules of Special Relativity and allow us to properly describe physical quantities such as position, velocity, and electromagnetic fields. Einstein showed that equations written in tensors, in four-dimensional curved spacetime (what we call *properly written*), are valid for all observers regardless of their velocity, acceleration, or location in gravitational fields. One property of the tensor equation shown above is that it represents 16 separate, but related, equations, thus the plural form of the name Einstein's Field Equations.

General Relativity is considered by almost all physicists to be the most profound and most beautiful of all physical theories. Brian Greene, author of *The Elegant Universe*, eloquently called it "The choreography of the cosmic dance of space, time, matter and energy." Paul Dirac said General Relativity was probably the greatest scientific discovery ever made. J.J. Thomson, Max Born, and Sir Arthur Eddington all went even further and said General Relativity was perhaps the greatest achievement of human thought. All these people were experts on good science.

Thomson discovered the electron and received the 1906 Nobel Prize in Physics. Dirac received the 1933 Nobel Prize and Born received the 1954 Nobel Prize, for their contributions to the development of Quantum Mechanics. Eddington received the Royal Society Royal Medal, the Royal Astronomical Society Gold Medal, and a knighthood.

I would say General Relativity is a pinnacle achievement of human culture to be cherished along with our greatest achievements in music, literature, art, medicine, and other fields of intellectual endeavor.

Einstein was once asked why he had been able to make his great discoveries while others had not. He replied, "It's not that I'm so smart, it's just that I stay with problems longer." Einstein was certainly wrong about the first part, but right about the second. He worked on some problems, almost continuously, for a decade or more. Such intensive thought is truly remarkable, particularly in our age when instant gratification doesn't come soon enough.

Conceptually, General Relativity is extraordinarily elegant, but its mathematics is extremely difficult. In nearly 100 years, only a handful of exact solutions to Einstein's Field Equations have been discovered. In the next chapter, we will discuss one of the most important solutions and find out what that tells us about curved spacetime near a massive body, such as a star or a black hole.

NOTES

[1] On a sphere, triangles covering a large fraction of the sphere's surface have internal angle sums of much more than 180 degrees, while triangles covering a small area have internal angle sums only slightly over 180 degrees. Consider a very small equilateral triangle. Each of its interior angles is nearly 60 degrees, while its external angles are nearly 300 degrees. The outside of the small triangle is also a triangle; it covers nearly the entire sphere, has the same three sides, and its internal angles are each 300 degrees, summing to 900 degrees.

21

Solving Einstein's Field Equations

In 1915, Einstein published his Theory of General Relativity with his Field Equations $G=8\pi T$. These equations are so complex, that initially, neither Einstein nor anyone else was able to solve them exactly. Even leading European mathematicians such as David Hilbert did not find an exact solution.

Einstein was able to derive a few approximate solutions for situations in which spacetime curvature was very small. This is a common approach when equations are too difficult to solve exactly. One tries to solve them approximately by assuming the complex part is small. Being able to calculate approximate solutions and knowing how to use them effectively is an essential part of the art of science.

Let's try a simple example: find the square root of 104. The square root of 100 is 10. Since 104 is not much larger than 100, the answer we seek is probably just a little bit larger than 10. Write the answer as $10+x$, where x is the "just a little bit." We want $104=(10+x)^2=100+20x+x^2$. Here is the key step: since x is small, x^2 is very small, and we will ignore it. The equation is now simpler: $104=100+20x$, hence $x=0.2$, and our approximate answer for the square root of 104 is 10.2. In fact, that's quite close to the exact answer that's a bit more than 10.198. (Our decision to ignore x^2 worked well because, in this case, $x^2=0.04$ and that's much smaller than what it's being added to, which is more than 100.)

Figure 21.1 Karl Schwarzschild (1873–1916) found the first, and perhaps most important, solution of Einstein's Field Equations. He died shortly thereafter, while serving as an artillery officer in World War I.

Getting even an approximate solution to the Field Equations is much more difficult, but hopefully this example illustrates the idea.

Einstein's approximate solutions are adequate to compute the bending of starlight by the Sun and the precession of the orbit of Mercury. This is because, as we will learn later, both are very small perturbations, no more than a few parts in a million. But what about situations where curvature is not a small perturbation, but instead leads to major changes? For those, nothing beats an exact solution.

German physicist and astronomer Karl Schwarzschild, shown in figure 21.1, set out to find an exact solution of $G=8\pi T$. Schwarzschild was a child prodigy, and had published a paper on celestial mechanics when he was only 16. Shortly before his death, Schwarzschild found the first, the most famous, and perhaps the most important solution of Einstein's Field Equations—a solution that will forever bear his name.

Before we get to it, let's make a new friend: *2M/r*. Yes, it's math, but it's not so bad is it? If you can remember *2M/r*, you will understand a great deal about spacetime curvature, black holes, and more. Our new friend tells us the curvature of spacetime at a distance *r* from an object of mass *M*. Here is Schwarzschild's exact solution of $G=8\pi T$:

$$ds^2 = -dt^2\,(1{-}2M/r) + dr^2\,/\,(1{-}2M/r) + dW^2$$

Wow! If it were simple, don't you think Einstein would have gotten it? OK, you can breathe again. Remember, the only part we need to deal with is our friend *2M/r*; it contains all the curvature. The curvature in the *dt²* term means time is curved, or warped as some say, and the curvature in the *dr²* term means space is also curved. Very far from the mass *M*, *r* is very large, *2M/r* is very small, and the curvature is nearly zero.

How significant is the curvature due to *2M/r*? Let's consider just the curvature of time, because it is generally more important than the curvature of space [1]. Time curvature slows the flow of time—where *2M/r* is large, clocks run slower than where *2M/r* is small. This is called *gravitational time dilation*. For example, on Earth's surface, time slows down by 4 seconds per century relative to somewhere very far from all massive bodies. Since all our clocks run at the same slower rate, we don't notice this effect and we think we are all "on time."

Global Positioning System (GPS) satellites orbit 12,000 miles above Earth. They are 4 times farther from Earth's center than we are on the surface, thus their r is 4 times larger and their clocks lose only 1 second per century. Since that differs from our clocks, GPS must be corrected according to General Relativity, otherwise it would drift by 6 miles per day. GPS readings that are correct now, would be off by 6 miles tomorrow, 12 miles the next day, etc. GPS would be useless without Einstein.

The strongest gravity in our solar system is on the surface of the Sun, where *2M/r* slows time by 222 minutes per century, about 4 parts in a million. Surprisingly, the Andromeda Galaxy, which we'll discuss in chapter 28, slows our clocks on Earth by 6 minutes per century. Even though it is 15 million, million, million miles away, its enormous mass more than compensates for that great distance. There may not be anywhere in the universe that is so remote that spacetime curvature does not have a significant influence. We are all connected by gravity.

On the *event horizon* of a black hole, time stops completely because *2M/r=1* (the curvature is 100%). Nothing ever changes on the event horizon of a black hole. There, the concept of time is meaningless.

We will discuss black holes in a chapter coming soon; exactly how soon depends on your spacetime curvature.

I have a few more words of explanation about Schwarzchild's equation. Those allergic to math might wish to skip to chapter 22.

Congratulations on bravely continuing. Everything you need to know about spacetime curvature can be derived from an equation called the *metric*. The metric lets you compute the distance between nearby points in four-dimensional spacetime. Schwarzchild's equation is the metric for curved spacetime outside a symmetric body of mass *M*. A symmetric body is perfectly round and does not rotate. Rotation selects one direction, the axis, as different from all others; hence a rotating object is not fully symmetric. This equation is valid only outside the massive body; thus *r* must be greater than the body's radius. In the above equation, $dW^2 = r^2\,da^2 + r^2\,db^2\,sin^2a$ is the square of the distance between two points on the surface of a sphere. There is no curvature in this term. In Schwarzchild's equation ds^2 is called the *interval* and is the square of the four-dimensional distance between nearby points separated in time by *dt*, in radius by *dr*, in latitude by *da*, and in longitude by *db*. The curvature is only in time and in the radial direction that points out from the body's center.

Schwarzchild's metric, but not the actual geometry of spacetime, becomes mathematically undefined at the *Schwarzchild radius r=2M*. This is because the denominator of the radial term becomes zero. This problem can be avoided by the use of other coordinate systems.

NOTES

[1] Space curvature is generally less important than time curvature because dr^2 is generally much less than dt^2. This is because $(dr/dt)^2$ is never more than v^2 and an object's velocity *v* is generally much less than 1, the speed of light.

22

General Relativity in Action

The bending of starlight by the Sun, discussed in chapter 19, was the most famous and most convincing early confirmation of Einstein's Theory of General Relativity. This chapter presents more recent examples of the effects of General Relativity and more precise tests of this great theory.

GRAVITATIONAL LENSING

After the initial sensation of the bending of starlight subsided, people found ways to use this phenomenon for practical advantage. Bending light is what makes telescopes work. Since the invention of the telescope 400 years ago, we have learned how to precisely shape glass to achieve extraordinary lenses and mirrors. General Relativity tells us that the gravity of massive celestial bodies bends light as well. Clearly, we cannot manipulate the shapes of stars and galaxies; we have to settle for what is out there. Nonetheless, *gravitational lensing* by massive cosmic bodies has become a valuable tool for astronomers, as we shall see.

Figure 22.1 shows an image of the Einstein Cross. Light from a very distant *quasar* is *lensed* by the gravity of a galaxy that happens to lie almost exactly between us and the quasar. Quasars are galaxies with massive black holes that voraciously consume vast amounts of matter and

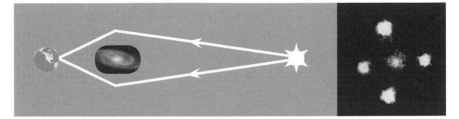

Figure 22.1. Light from a very distance quasar is bent by the gravitational lensing of a closer galaxy. Four images of the quasar are focused onto Earth producing the image on the right, the Einstein Cross. The central spot is light from the lensing galaxy. Image courtesy of NASA Hubble.

emit spectacular amounts of light (more on this in chapter 26). Quasars can be a million, million times brighter than our Sun, and can be visible across virtually the entire expanse of the universe. When the universe was much younger, quasars were in their heyday. As the universe matured, galaxies became more sedate and quasars became rare. This quasar's light has been traveling toward us for 10 billion years. The *lensing* galaxy is much closer, only 500 million light-years away. The quasar's light is bent by the gravity of the closer galaxy and some of it is focused onto Earth. Galaxies have a wide variety of shapes and are almost never ideal lenses. This galaxy focuses four separate images of the distant quasar in our direction. Thus figure 22.1 has the image of the closer galaxy in the center surrounded by four images of the more distant quasar. We know these are four images of the same object because they all have the same distinctive light spectra. The Einstein Cross was discovered by John Huchra in 1985, 30 years after Einstein's death and was named in his honor.

Color Plate 11 shows examples of *Einstein rings* that also result from gravitational lensing of light from distant sources by nearer massive bodies. In the upper right, the fortuitous alignment of a distant source, a lensing galaxy cluster, and Earth produces an almost perfectly circular, blue Einstein ring. Usually, the alignment is less ideal, resulting in partial rings. Numerous partial rings are seen on the left, lensed by galaxy cluster Abell 2218 that is 3 billion light-years away. The boxed region is magnified in the lower right image. This image is a dramatic demonstration of astronomers taking maximum advantage of what nature provides.

The arrow added to the image points to the farthest and oldest galaxy yet seen. OK, it's only a small red blob with no majestic spiral arms. That's what you have to expect from the farthest galaxy yet seen—if this one were easy to see, there would probably be another even farther away that was barely visible. Modest as it is, this is the farthest and the oldest object in the universe that you will ever see, at least until you read chapter 34. Light from the galaxy seen here has been traveling toward us for about 13 billion years, 94% of the age of the universe. Gravitational lensing increased the brightness of its image 30-fold. Without lensing, this ancient galaxy would be too dim to detect with current telescopes. That would be a shame; after such a long journey, its light deserves to be observed and celebrated.

PRECESSION OF PLANET MERCURY

General Relativity makes an important prediction about the motion of planets in our solar system. In Newton's theory of gravity, a single planet orbiting a star travels along an elliptical path, as shown on the left side of figure 22.2. The planet always orbits on the same path; the position of the ellipse never changes. This is because Newton's gravitational force varies as $1/r^2$; it is an *inverse-squared* law. In General Relativity, the force is more complex; it is modified by our old friend $2M/r$. This introduces an additional $1/r^3$ term that causes the ellipse itself to rotate around the Sun—the ellipse *precesses*. The precession rate is very small and can be adequately measured only for Mercury, the planet closest to the Sun and most strongly affected by its gravity. General Relativity predicts that Mercury's ellipse rotates 30 millionths of a degree per orbit, rotating one full turn every 12 million orbits of Mercury, which takes 3 million Earth-years. Astronomers had observed the precession of Mercury's orbit before Einstein. Part of the precession is due to the other planets pulling on Mercury. After adjusting for the other planets, a residual precession remains that Newton's theory of gravity cannot explain, an "anomalous" precession. This anomalous part has been measured to a precision of 1% and it exactly matches Einstein's prediction.

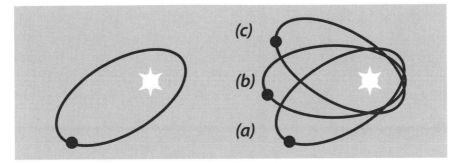

Figure 22.2. Orbit of Mercury according to Newton (left) and Einstein (right). In Newton's theory of gravity, a single planet (black dot) moves along an elliptical path around the Sun, but the ellipse remains fixed. Einstein said that the ellipse precesses—it slowly rotates around the Sun from (a) to (b) to (c), making one full turn every 12 million orbits of Mercury.

GRAVITATIONAL REDSHIFT

Einstein's theory of gravity also says that light loses energy as it climbs out of a gravitational field. As a pendulum swings up, its kinetic energy is converted to gravitational potential energy and the pendulum slows down. Since light has energy, and energy is equivalent to mass, Einstein reasoned that light moving upward against gravity must also lose energy. Light cannot slow down—it always moves at the same speed—but it can lose energy by increasing its wavelength and decreasing its frequency. For visible light, this means shifting toward the red end of the spectrum. This is called *gravitational redshift*. A series of remarkable experiments, done by Harvard physicists Pound, Rebka, and Snider, confirm Einstein's prediction within the observational precision of 1%. To achieve this precision, they had to measure the frequency of light to ±5 parts in a million, billion—an uncertainty of only 5 in the 15TH decimal digit. The extraordinary efforts physicists expend to test General Relativity attest to the extraordinary importance of Einstein's theory.

General Relativity also predicts an interesting new phenomenon called *gravity waves* that are ripples in the fabric of spacetime caused by the motions of massive objects. A beautiful confirmation of Einstein's prediction of gravity waves is discussed in chapter 24.

23

White Dwarfs

When the core of a dying star collapses, a phenomenal amount of gravitational potential energy is released, perhaps as much energy as the star released in its entire prior life. The star's outer layers are blown away and form what astronomers call a *planetary nebula*. These nebulae are truly beautiful, but they have nothing to do with planets—they are star dust. The star's collapsed core becomes one of three possible, extremely dense, exotic objects: a white dwarf, a neutron star, or a black hole. The latter two are the subjects of the next three chapters.

This chapter is about *white dwarfs*. Some famous planetary nebulae surrounding white dwarfs are shown in figure 23.1 and in Color Plates 12, 13, and 14. In each of these images, the collapsed core—the white dwarf—is a small white dot at the center of the nebula. All these images were taken by NASA's Hubble Space Telescope.

Figure 23.1 shows the planetary nebula NGC 2440, which is 4000 light-years away. The white dwarf at its center has one of the highest stellar surface temperatures ever measured: 360,000 °F.

Color Plate 12 shows the Cat's Eye Nebula taken in three different wavelengths: nitrogen atoms are shown in green, hydrogen atoms in red, and oxygen atoms in blue. The Cat's Eye is 3000 light-years away and is the remnant of a star that died 1000 years ago. The complexity of the concentric gas shells, high-speed jets, and shock waves makes the Cat's

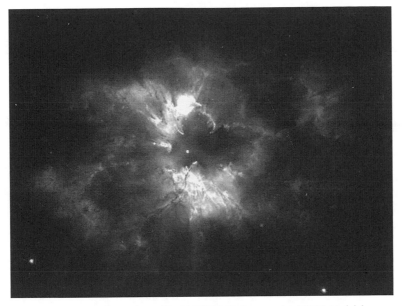

Figure 23.1. White dwarf in NGC 2440. Image by NASA Hubble.

Eye Nebula a remarkable sight and the subject of many lively scientific discussions.

Color Plate 13 shows the Spirograph Nebula, which is 2000 light-years away. Its diameter is ⅓ of a light-year and while that is "only" 2 trillion miles, it is still 300 times larger than our solar system. The now-dead star that created this nebula was probably very much like our Sun; thus this is what our solar system may look like 5 billion years from now.

Color Plate 14 shows the Eskimo Nebula, which is 5000 light-years away and 8 light-years across. It is named for its resemblance to a person wearing a fur parka. A *jet* of material from the center of the Eskimo Nebula is coming toward us at a speed of over 400,000 mph. (Not to worry—the jet will take 9 million years to get here and will dissipate by then.) The Eskimo Nebula is 1000 years old.

Just as a star's mass is the key factor in determining its life and death, it also determines its afterlife. The lightest stars become white dwarfs. About 95% of all stars eventually produce a collapsed core with a mass below 1.4 *Msun*; all these become white dwarfs. (Recall *Msun* is our abbreviation for the mass of our Sun.)

Since our Sun is below this mass limit, after it passes through the red giant phase, its core will become a white dwarf and its outer layers will likely produce a planetary nebula. Collapsed cores with masses above about 3 *Msun* become black holes. This category includes only the real heavyweights, the most massive 1/1000TH of all stars. The remaining 5% are middling-mass stars; they eventually become neutron stars.

White dwarfs are composed of normal stellar material—electrons and atomic nuclei. But all this matter is compressed to an astonishing degree. When nuclear fusion ends, and the pressure it created wanes, gravity crushes a star about the size of our Sun down to something only about the size of Earth. Its volume becomes a million times less, and its density becomes a million times greater. A teaspoon of white dwarf can have a mass of 30 tons. It would take a king-sized bulldozer to push around that teaspoon. Gravity on its surface is a million times Earth's gravity and would crush even the strongest steel into a flat sheet, one atom thick. This is not a place you want to visit.

Initially, white dwarfs are very hot, heated by the enormous compression they have suffered. But because they have exhausted their nuclear fuel, they have no continuing source of energy. They slowly cool and fade away over billions, or even trillions, of years. Finally, they are merely dead embers floating in space. At least that's what happens if they are unmarried.

About a third of all stars are single—alone in space like our Sun. But most stars are in couples or threesomes, and a few are in even larger multiples. Couples are called **binary systems**. Stars with partners generally form at the same time and have the same age. But they may have very different masses, and therefore have very different lifetimes and ultimate fates.

Some binary systems orbit quite far apart, but the most interesting are **close binaries** in which the heavier star is slightly below the maximum mass for white dwarfs. Let's say the heavier star produces a collapsed core with a mass of 1.2 *Msun*. As the heavier star of the couple, it progresses through the stellar life cycle faster than its partner. It becomes a white dwarf while its lighter partner is still burning hydrogen. When the lighter partner becomes a red giant and expands, perhaps 100-fold,

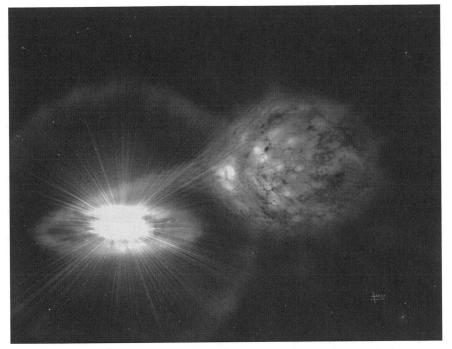

Figure 23.2. Illustration shows a white dwarf on the left cannibalizing a companion star on the right. © David A. Hardy www.astroart.org/STFC

it may become the victim of stellar cannibalism. The gravitational force near a white dwarf is much greater than at the surface of a red giant. If the white dwarf is close enough, it may pull in gas from its partner's extended surface. It gains mass at its partner's expense, as illustrated in figure 23.2.

TYPE IA SUPERNOVAE

The most exciting part comes when the white dwarf's mass approaches 1.4 *Msun*. The increased mass triggers nuclear fusion reactions that weren't possible before. As discussed in chapter 18, in normal stars, the nuclear fusion rate is regulated by a delicate balance of gravity and pressure. However, in a white dwarf, gravity has overwhelmed normal gas pressure.

With nothing to regulate them, the new fusion reactions run unchecked and the star explodes catastrophically in a *Type Ia supernova* (Ia is pronounced "one-A"). It releases a stupendous amount of energy and the star blows itself to pieces, totally disintegrating. Type Ia supernovae are among the most violent and spectacular events in the cosmos.

Type Ia supernovae have some distinctive, identifying features, including their light spectra and the evolution of their energy output. Their spectra strangely contain almost no evidence of hydrogen. Recall from chapter 16 that each element has its own unique light emission and absorption frequencies that allow astronomers to determine the chemical compositions of celestial light sources. The absence of hydrogen is very unusual because it is by far the most common element in the universe. Its absence in Type Ia supernovae indicates that any hydrogen that the star did not convert to helium must have been blown away into its planetary nebula during the core's collapse to a white dwarf. That would have occurred well before the white dwarf became a supernova. The evolution over time of the energy output of Type Ia supernovae is also distinctive. Their luminosity grows rapidly, peaks in about two weeks, and gradually declines. The show is over in a few months.

Their unique characteristics allow astronomers to reliably distinguish Type Ia supernovae from other spectacular cosmic events. And their immense brightness allows us to see them even at vast distances.

The best thing about Type Ia supernovae is that they all emit almost exactly the same amount of energy. This is because they all occur in the same way—a white dwarf slowly reaching a specific mass. This makes them our best *standard candles*—objects with a known energy output. By measuring how much of its light reaches Earth, astronomers can determine how far away a Type Ia supernova is. This makes Type Ia supernovae a great way to measure long-range astronomical distances, which is important because measuring distances is one of the most difficult tasks in astronomy. In chapter 33, we will discuss further the use of Type Ia supernovae in determining cosmic distances and in measuring the expansion of the universe.

DISTANCE TO SN1987a

Supernovae can also provide another means of measuring astronomical distances. An important discovery was made several months after the explosion of SN1987a that allowed astronomers to measure its distance very precisely. Before the star exploded, it ejected some of its outer layers. As shown in figure 23.3, some of this material formed rings around the star. The brightness of the inner ring increased dramatically 245 days after the supernova occurred—when it was hit by the supernova's immense blast of light and neutrinos. (These neutrinos are so energetic that they travel at virtually the speed of light.) Since we know how long it took light to reach the ring, and since we know the speed of light, we know the ring's diameter was 1.34 light-years. From that and the angle subtended by the ring, simple geometry yields the distance to SN1987a: 168,000 light-years with a precision of 2%. This is superbly accurate and greatly enhances the precision with which we know the distance to SN1987a's host galaxy, the Large Magellanic Cloud.

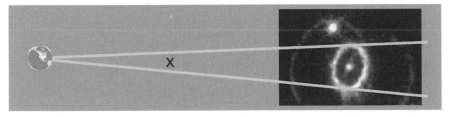

Figure 23.3. A precise measurement of the distance to supernova SN1987a was obtained when a ring around the star lit up 245 days after the supernova was first seen. From the ring diameter and the observed angle x subtended by the ring, the distance was found to be 168,000 light-years with a precision of ±2%. Image by NASA Hubble.

24

Neutron Stars

Stars whose collapsed cores have masses above 1.4 *Msun* are too heavy to be white dwarfs. This maximum mass for white dwarfs is called the *Chandrasekhar limit*, in honor of Indian-American astrophysicist Subrahmanyan Chandrasekhar. "Chandra" was born and educated in India. After his college graduation in 1930, he received a scholarship to study at Cambridge. It was a long, boring boat ride to England (this was before the "Love Boat"). During the voyage, Chandra calculated the maximum mass a white dwarf could have. Everyone else in the world thought there was no maximum, but Chandra showed there is a limit to the gravitational force that normal matter can withstand. If a star is too massive, he said, it would collapse to almost nothing.

To Einstein, Eddington, and all who followed their lead (which was then virtually everyone), the idea of stars collapsing to nothing was simply unacceptable. They felt nature would not tolerate something so "ugly." Chandra got no love in Europe; in fact, he was ridiculed for proposing something so "unnatural." He moved to the University of Chicago in 1937. Eventually, physicists realized Chandra had been right. He received the 1985 Nobel Prize in Physics, and NASA named its x-ray space telescope in his honor. Figure 24.1 shows Chandra and his namesake.

Chandra discovered that under an extreme gravitational force, electrons are squeezed into and absorbed by nuclei. Electrons combine with

Figure 24.1. Subrahmanyan Chandrasekhar (1910–1995) and the x-ray space telescope named in his honor, shown in the cargo bay of the Space Shuttle awaiting launch in 1999. Images by NASA.

protons to form neutrons, and in the process they emit neutrinos. Since a neutron is more massive than an electron-proton pair, this reaction requires a substantial energy source. That source is gravity. Even in the super-compressed matter of a white dwarf, an electron's wavelength is a thousand times larger than a neutron's. Getting rid of electrons allows the star's core to collapse to something a thousand times smaller still—a *neutron star.*

The core collapses into a neutron star in as little as 1 second, creating another type of supernova. This is called a *core collapse supernova,* as opposed to a Type Ia supernova that results from the disintegration of a white dwarf. The phenomenal energy released by a core collapse supernova is distributed in several ways, including: creating neutrons; radiating light; and providing the kinetic energy to blast the star's outer layers off into space. But the vast majority of the energy goes into a stupendous burst of neutrinos. Neutrinos may carry away 100 times more energy than does light. When SN1987a exploded, its neutrinos were detected on Earth, a million, million, million miles away. Several very large neutrino detectors had been built to study solar neutrinos and were capturing a few neutrinos per day. SN1987a's neutrino blast lit up these experiments like Christmas trees, but only for 10 seconds.

In a neutron star, almost all the electrons and protons combine to form neutrons and almost all the space between nuclei disappears. The star is effectively one giant nucleus composed almost entirely of neutrons. Without the "larger" and lighter electrons, the density of matter is incredible. A star more massive than the Sun collapses to only 15 miles or less in diameter. A teaspoon of neutron star has a mass of billions of tons. Gravity on its surface is 100 billion times Earth's gravity. The curvature of spacetime from neutron stars is exceeded only by that from black holes. On a neutron star *2M/r* can equal ⅓—a clock (if it weren't crushed) would lose eight hours each day.

Rotating objects must spin faster if they shrink. Our Sun rotates around its axis once every 28 Earth-days. Since neutron stars are vastly smaller, they rotate much faster. Most rotate many times per second—some rotate hundreds of times per second.

PULSARS

When normal stars collapse into neutrons stars, their magnetic fields are also compressed and become enormously intense, as much as a trillion times Earth's magnetic field. Some neutron stars have their magnetic poles oriented in a different direction than their rotational axes. So does Earth; our magnetic north is different from true north. The combination of skewed magnetic axis, fast rotation, and immense magnetic fields puts some neutron stars in a special class called *pulsars*. Pulsars sweep radio waves across the heavens, much like lighthouse beacons sweep across our sky. If we're lucky, their beams sweep past us. By detecting pulses of radiation each time its beam goes by, astronomers can measure a pulsar's rotation rate. Figure 24.2 shows an x-ray motion picture of the pulsar at the center of the Crab Nebula taken at 1000 frames per second. These images show the Crab's pulsar rotating 33 times per second.

Pulsars are among the most stable "clocks" in the universe. Their pulse rates can keep time to better than ±1 second in a million years. In 1974, American radio-astronomers Joseph Taylor and Russell Hulse found a new type of object in the heavens. Their great discovery was the

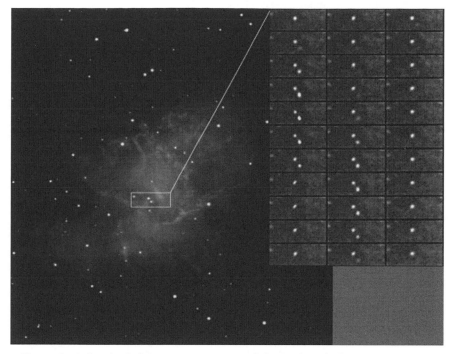

Figure 24.2. On the left is an x-ray image of the Crab Nebula. On the right is an x-ray motion picture of the Crab's center taken at 1000 frames per second. As it spins 33 times per second, the Crab's pulsar comes into and goes out of view near a fixed background star. Image by NASA Chandra.

first **binary pulsar**. As shown in figure 24.3, the binary pulsar consists of two neutrons stars orbiting one another in under 8 hours—their "year" lasts less than 8 Earth-hours. One of the pair is a pulsar that is rotating 16 times per second. Its pulse rate is extraordinarily stable, allowing Hulse and Taylor to precisely measure the stars' orbital properties.

GRAVITY WAVES

Einstein said that gravitational changes ripple through the fabric of spacetime as **gravity waves** traveling at the speed of light. As massive neutron stars rapidly orbit, they flex spacetime back and forth, emitting gravity waves that carry off energy, as illustrated in figure 24.4.

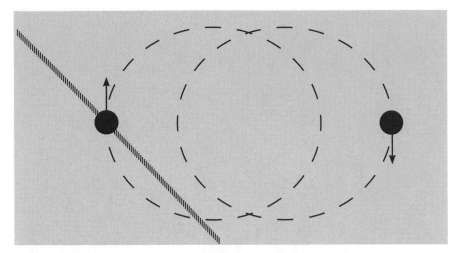

Figure 24.3. Above schematic of the Hulse-Taylor binary pulsar shows two neutrons stars orbiting one another (arrows indicate their velocities). The left star is also a pulsar; its beams of radiation and high-energy particles are represented by the crisscrossed lines.

Because total energy must be conserved, the stars' orbital energies must gradually decrease to power the gravity waves. As their orbital energy drops, the stars move closer to one another, and their "year" shortens. The stars are expected to collide in about 300 million Earth-years. Their merger will probably create a black hole.

Given the measured masses, rotation rate, and orbital diameter, General Relativity predicts that their "year" is shortening by 75.8 seconds per million Earth-years. Hulse and Taylor measured the shortening of the orbital period to be 76.5 ± 0.8 seconds per million Earth-years, where 0.8 is the uncertainty due to the limitations of their instrumental precision. In English, this means the experiment says the rate is most probably between 75.7 and 77.3, which confirms the prediction of General Relativity to a precision of 1%.

For their discovery and their remarkable observations, Hulse and Taylor shared the 1993 Nobel Prize in Physics.

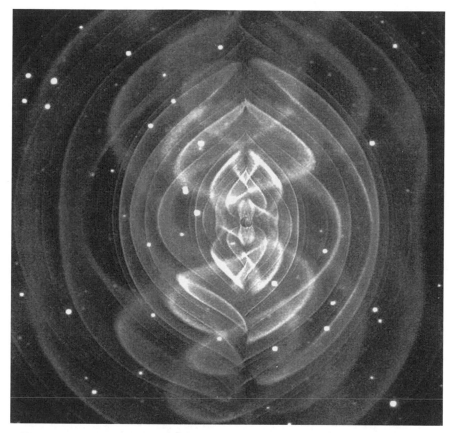

Figure 24.4. Computer simulation of gravity waves rippling through the fabric of spacetime due to two massive celestial bodies in a close orbit around one another. Image courtesy of NASA, Caltech.

Isn't it amazing that Einstein could sit at his desk with only paper and pen, and precisely predict the behavior of a phenomenon that had never been seen or even imagined, and that is 100 million, billion miles away? This is a stunning example of the incredible potential of the human mind. Granted, few of us will match Einstein, but each of our DNA is 99.9% the same as his.

Imagine what extraordinary achievements you are capable of!

What are Black Holes?

However strange white dwarfs and neutron stars may seem, black holes are even more bizarre.

If the collapsing core of a dead star is massive enough, no force of nature can resist its crushing gravity. The maximum mass that a neutron star can support is approximately 3 *Msun* (the exact value is still uncertain). Any collapsing core with a greater mass must become a black hole. In less than a heartbeat, the entire core collapses from the size of a neutron star down to almost zero size. A 6 *Msun* core collapses from 12 miles across to virtually nothing in about $1/10,000^{\text{TH}}$ of a second.

SINGULARITY AND PLANCK LENGTH

How small is "virtually nothing"? It's as small as anything can be in our universe, probably less than a trillionth of a trillionth of the size of the smallest atom. While no one knows for sure, most physicists believe there is a smallest possible size that can exist in our physical universe, as we will discuss in chapters 28 and 39. That smallest size may be 1.6×10^{-35} meters, which is called the *Planck length*. Words are not adequate to describe how small this is. At this scale, our concepts of space and time are probably invalid. Space and time may become quantized, somewhat like electrons' energies are quantized in an atom. The Planck length may be the size of one quantum of space. If space really is quantized, that would be the only thing that stops a black hole from collapsing to zero size. It would have to stop at one quantum if nothing smaller exists.

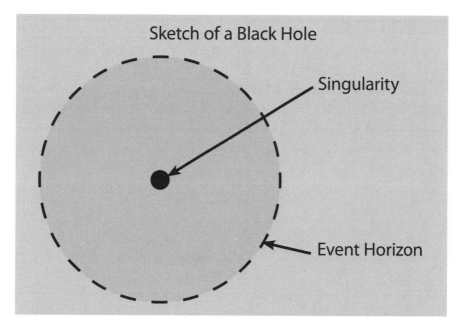

Figure 25.1. Every black hole has a central singularity that is surrounded by an event horizon. The singularity is vastly smaller than an atom; nonetheless, it contains all the black hole's mass. The event horizon is the limit of what we can see from the outside, and it is where the escape velocity equals the speed of light.

All of a black hole's mass is compressed into an unimaginably small volume called a ***singularity***, illustrated in figure 25.1. How can so much matter be compressed into almost nothing? Recall our discussion in chapter 16. Particles have a wavelength but their intrinsic size is zero; there is nothing "solid" inside them. Their wavelengths become smaller as their energies increases. The tremendous energy liberated by a star's gravitational collapse is more than enough to squeeze the wavelength of every particle in the star down to almost nothing.

EVENT HORIZON AND ESCAPE VELOCITY

Every black hole is surrounded by an ***event horizon***. This is not a physical object; it's the limit of what we can see and the point of no return. Everything that happens inside the event horizon is beyond the view of everyone outside. We are accustomed to a different sort of horizon here on Earth: for example, when we look out over the ocean. On Earth, the horizon is also not a material object, but rather the limit of how far we can see. Some once thought the horizon was the end of the Earth, but, of course, we now know that's not true. Earth's horizon is due to the curvature of Earth's surface, and as the observer moves, the horizon moves as well. (Heard the joke about the photographer who wanted to move closer to the horizon to get a better shot?) A black hole's horizon is due to the curvature of spacetime, and it remains in a fixed position surrounding the singularity. It's best not to get too close for that great photo.

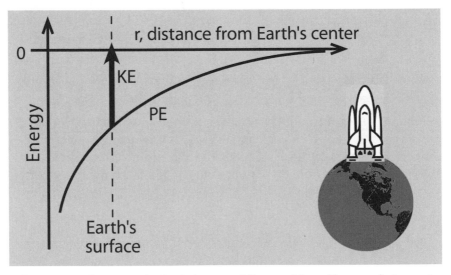

Figure 25.2. Escape velocity is how rapidly an object, like a rocket, must move to escape the gravitational pull of a large body of mass *M*. The rocket's potential energy *PE*, represented by the curved line, is negative and varies with the distance *r* from the center of mass. To escape, the rocket's kinetic energy *KE* (dark arrow) must raise its total energy to at least zero. From the equations for *PE* and *KE*, we find that the escape velocity is $v^2 = 2M/r$.

The event horizon of a black hole is where the *escape velocity* equals the speed of light. That merits an explanation. Figure 25.2 shows a rocket ready to blast off Earth's surface. The rocket starts with negative gravitational potential energy, negative because it takes energy to lift the rocket out of Earth's gravitational field. To escape Earth's gravity, the rocket must be able to move very far away to where the potential energy is essentially zero. To do that, the rocket must have enough kinetic energy (high enough velocity) to raise its total energy (kinetic plus potential) up to zero or more [1]. This means the minimum escape velocity v is given by: $v^2 = 2M/r$ (our old friend $2M/r$ again). Any object in a gravitational field can escape only if it can move away from the center of gravity with at least this velocity.

Now, consider an intrepid astronaut who decides to take a space walk and explore the inside of a black hole. We will stay in the mother ship, far from danger, and watch what happens. Far from the event horizon, the escape velocity might be *0.1c*, 10% of the speed of light. See figure 25.3. Here, the astronaut can escape to safety if his **NASA** jet pack can propel him at 67 million mph (the manual says that's A-OK). As the astronaut moves closer, r decreases, and the escape velocity increases. At some point, the escape velocity is *0.9c*, or 604 million mph (that's tempting fate). When the escape velocity is *1.0c*, equal to the speed of light, the astronaut is at the event horizon, with no chance of ever coming back.

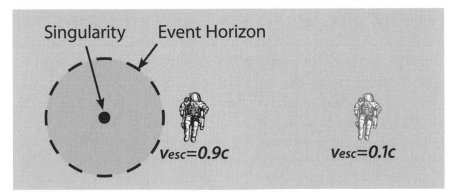

Figure 25.3. As an astronaut approaches a black hole, his escape velocity increases and reaches the speed of light at the event horizon.

Astronauts approaching from different directions would reach different points in space where the escape velocity equals the speed of light. The event horizon is the collection of all such points. If a black hole is not rotating, its event horizon is a spherical surface surrounding the singularity with radius $r=2M$. A black hole with a mass of 10 *Msun* has an event horizon radius of 20 miles. The most massive black holes we know have masses one billion times larger and their event horizon radii are 20 billion miles, 200 times the radius of Earth's orbit.

Inside the event horizon, $2M/r > 1$, thus the escape velocity exceeds the speed of light and nothing, not even light, can move fast enough to escape. Here, all roads are expressways to the singularity. Just as everything outside the event horizon must move inexorably forward in time, everything inside the event horizon moves inexorably toward the center. Objects inside can no more resist moving toward the singularity than we can resist getting older.

The name "black hole" was coined by American physicist John Archibald Wheeler of Princeton. He called it "black" because nothing comes out, not even light; hence, it appears black. He called it a "hole" because once anything enters the event horizon it is gone forever, as if it fell out of our universe into an unobservable, bottomless abyss.

It turns out our astronaut is a physicist (who else?) intent on making careful observations and discovering new science, hoping to win fame, glory, and a Nobel Prize. His spacesuit is rigged with biometric sensors that continually transmit his vital signs back to the mother ship. His plan has just two problems. Firstly, the Nobel Prize is not awarded posthumously. Secondly, as Caltech's eminent blackhologist Professor Kip Thorne said, once our physicist crosses the event horizon he suffers a fate even worse than death: the inability to publish.

As our astronaut approaches the black hole, what does he see? Our physicist astronaut sees himself falling faster and faster toward a black void. For him, the event horizon is nothing special, just another place in endless black space. He blows on by without even noticing. But once inside, he can never again communicate with the world outside the event horizon because no signal can travel faster than light, and even light cannot move fast enough to escape the black hole's gravity.

GRAVITATIONAL TIDAL FORCES

At some point, exactly where depends on the black hole's mass, our astronaut starts feeling the black hole's growing *tidal forces*. These gravitational forces are like those from our Moon that cause the ocean tides on Earth, except a black hole's tidal forces are super-sized. Tidal forces pull his head toward the singularity and pull his feet away from it, **stretching him**. They also push in on his left, right, back, and front, squeezing him. Soon he becomes what we all want to be, but few are—tall and thin. Soon thereafter, he becomes *very tall* and *very thin*. Astrophysicists have a special scientific term for this: *spaghettification*. Eventually his body disintegrates, and even his atoms are torn apart. All this happens quickly. If the black hole's mass M is 1 million *Msun*, he would say (if he still could) that it took only 15 seconds to travel from the event horizon to the singularity 2 million miles away. If M were 80 *Msun*, he would take $1/800^{TH}$ of a second to travel the 160 miles to the singularity. In the end, his remains will forever reside within the incredibly small singularity. See figure 25.4.

THE END OF TIME

How does all this appear to us in the mother ship? Very different! At first, we see the same scenario—our astronaut falls faster and faster toward the black hole. But as he approaches the event horizon, we see him move slower and slower until he is hardly moving at all. Why? As we see it, his time runs slower and slower. We observe his watch slow down, his heart beat slower, and all the biochemistry in his body run slower. All this is explained by Einstein's Theory of General Relativity—in highly curved spacetime (in strong gravity) where $2M/r$ is large, time runs slower.

We also see the image of our astronaut getting progressively redder and progressively dimmer, as light reflecting from his body struggles to escape the black hole's immense gravity. Eventually, we see him come to a full stop at the event horizon. There time stops completely and nothing ever changes.

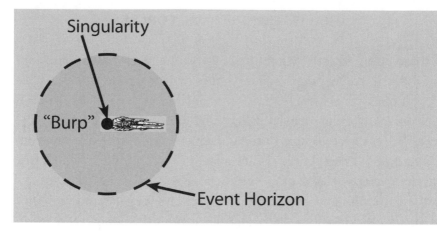

Figure 25.4. Astronaut suffers spaghettification from black hole's tidal forces and is absorbed into the singularity, never to return or be seen again.

While his image quickly becomes too red and too dim for us to see, in some sense his image will always remain on the event horizon along with the image of everything else that has ever "fallen in."

GRAVITY: BLACK HOLE vs. A STAR

A common misconception about black holes is that their immense gravity can suck up everything in the universe, or at least everything in a galaxy. While they do consume everything that enters their event horizons, it is important to realize that event horizons may be very small. The curvature of spacetime (the strength of gravity) due to any round object is *2M/r*. At the same distance *r*, the same mass *M* produces the same force regardless of whether it is a black hole, a normal star, or the fruitcakes of Christmas past.

In figure 25.5, the wide gray line tracks the acceleration due to the Sun's gravity at various distances from the Sun's center. The narrow black line is the acceleration of gravity due to a black hole of the same mass at the same distance. The details of how this is plotted are discussed in note [2]. Here, let's discuss what the figure means. Starting at Earth on the right side, the figure shows that the Sun and the black hole have the

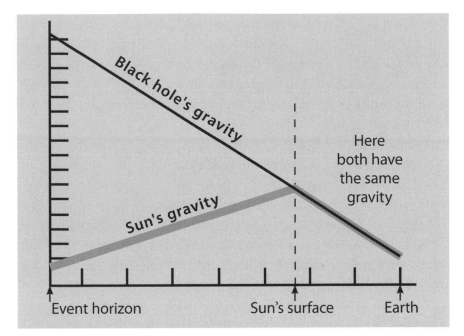

Figure 25.5. Acceleration of gravity (vertical axis) at various distances (horizontal axis) from the Sun's center due to the Sun (broad gray line) versus a black hole (thin black line) of the same mass. See note [2].

same gravitational strength. Moving left, toward the center of gravity, both strengths increase in lockstep. However, things change inside the Sun's surface. Inside the Sun? That's a strange place to go; it's not on the itinerary of anyone who wants to survive. But we can think about it without getting hot under the collar. Inside the Sun, its strength of gravity decreases [3]. But the black hole's gravity continues increasing, as before.

As the figure shows, a black hole's gravity is greater than the Sun's gravity only *inside* the Sun. At the black hole's event horizon, its gravity is 10 billion times more than the Sun's surface gravity. But this is only because we can get much closer to the center of a black hole and still be outside all its mass. Only inside the star's surface is a black hole's gravity stronger than the gravity of a star of the same mass.

While trip insurance is not available for either destination, it's safer to be 200,000 miles from the center of a 1-solar-mass black hole than the same distance from the center of the Sun (its radius is 435,000 miles).

Just as there are stable orbits around stars, there are also stable orbits around black holes. The closest stable orbit around a non-rotating black hole is at 3 times the radius of the event horizon, at $r=6M$. If our Sun were to become a black hole (there is no natural way we know that this could happen), Earth would continue to have a stable orbit exactly where it is now.

Another peculiar black hole phenomenon is their ability to bend light so strongly that light can circle completely around them, even several times. If a massive object falls into a black hole, some of its energy may be converted into a burst of radiation, an explosion of light called a *flare*. If the flare occurs just outside the event horizon, some photons may orbit the black hole once, some photons may orbit twice, and some may orbit many times as they spiral their way out and eventually escape. Black holes can be discovered and their masses measured from the timing of radiation detected from such flares. The radiation arrives in a series of bursts, with the same time delay between bursts. The first burst arrives after orbiting the black hole once, followed by a second burst that orbits twice, etc. These are called *light echoes*. Each echo corresponds to photons making a different number of orbits as they spiral their way out. The time between bursts is the same because the orbital circumferences are almost identical. The time difference tells us the size of the event horizon, which tells us the mass of the black hole. NASA is considering using this effect as a means to detect more black holes.

NOTES

[1] Gravitational potential energy equals $-Mm/r$, in natural units, where M is Earth's mass, m is the rocket's mass, and r is the rocket's distance from Earth's center. On Earth's surface the potential energy is negative. To escape Earth's gravity, the rocket must be able to move very far away, where its potential energy *increases* to zero. This requires kinetic energy $\frac{1}{2}mv^2$ sufficient to raise the rocket's total energy $\frac{1}{2}mv^2 - Mm/r$ to at least zero. As the rocket moves away, its kinetic energy

is converted to potential energy, while its total energy is unchanged. The minimum escape velocity is given by $v^2 = 2M/r$.

[2] Because figure 25.5 deals with values that vary over many factors of ten, it is plotted in a special way: each tick mark on the vertical and horizontal axes stands for a factor of ten. This means as one moves to the right 2 tick marks, the distance increases from 1 to 10 to 100, whereas in the usual plot it increases from 1 to 2 to 3. The vertical axis is similar. This is called a *log-log* plot. With a normal plot, I'd have to cut down a whole forest to get enough paper to make this plot.

[3] Gravity's strength increases rapidly as distance r decreases; the acceleration a is proportional to M/r^2. Outside a massive body like the Sun, when the distance r is cut in half, a increases 4-fold. But inside, M also changes. Anywhere inside a massive hollow shell, the force due to the shell's gravity is zero; the attraction of the mass in one direction is exactly cancelled by the attraction of the mass in the opposite direction. This is true in both Newton's and Einstein's theories of gravity. At any point within the Sun a distance r from the center, the Sun's mass can be divided into two parts: the interior is everything closer to the center than r, and the exterior is everything farther away. The exterior is a hollow shell; hence it exerts zero force at our point. Only the interior part of the Sun's mass contributes to the force of gravity at r. If the Sun's density were the same throughout, the amount of mass within r would be proportional to r^3. If r were cut in half, the amount of mass contributing to gravity would go down 8-fold. Even after General Relativity and Quantum Mechanics, 8 is still bigger than 4. Inside the Sun, a decreases as r decreases; if r is cut in half, a is also cut in half (up 4-fold due to $1/r^2$, down 8-fold due to M).

26

The Care and Feeding of Black Holes

It is likely that all existing black holes were formed from the collapse of truly massive stars. During their prime, these stars probably had masses of 30 or more *Msun*. Even after shedding their outer layers, they produced collapsed cores with masses exceeding 3 *Msun*. The gravity of such massive cores compressed them into black holes. But the story does not stop there.

BLACK HOLES CAN BECOME SUPER-MASSIVE

Black holes suck in everything that enters their event horizons. When a black hole is in the middle of the action, in the core of a galaxy, it can become super-massive by dining on a rich diet of gas, stars, and the odd planet. Its mass increases in direct proportion to what it consumes; if a star with mass 2 *Msun* falls into a black hole of mass 7 *Msun*, the black hole's mass increases to 9 *Msun*. As black holes gain mass, their event horizons grow larger but their singularities remain infinitesimal. Some black holes have grown to tip the scales at billions of solar masses. Figure 26.1 is adapted from a NASA chart based on Hubble Space Telescope data that shows a correlation between the ultimate mass of a black hole and the availability of matter to feed it. The trend line clearly shows that the

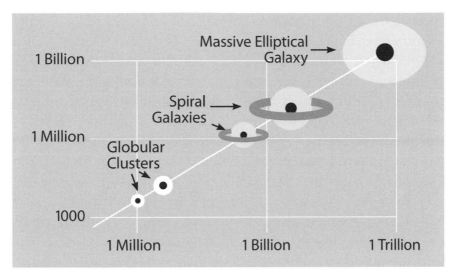

Figure 26.1. Adapted from a NASA chart based on Hubble images, plot shows that black hole masses (vertical axis) grow in proportion to the mass of nearby stars and gas (horizontal axis). Chart indicates black holes consume about half of 1% of their surroundings. Masses are in solar mass units.

more massive its surroundings the more material a black hole consumes and the more massive it becomes. These data indicate black holes consume about half of one percent of their neighborhoods. They are voracious, but at least so far, 99.5% of the surrounding material survives.

ACCRETION DISKS AND JETS

The right side of Color Plate 15 illustrates how a black hole can consume a companion star. If the companion is close enough, the black hole's gravitational pull on the companion's outer layers exceeds its own. Gas flows from the unfortunate companion toward the black hole. The in-falling material swirls around the black hole and forms an *accretion disk*. As gas nears the black hole, its gravitational potential energy converts into kinetic energy. The gas orbits ever faster and becomes ever hotter. The central black hole in galaxy M84 is estimated to have a mass of 300 million *Msun*; the interior of its accretion disk is orbiting at 880,000 mph.

At elevated temperatures, matter ionizes—electrons separate from atoms, leaving the atoms positively charged. This state of matter is called *plasma*. Material in an accretion disk heats to billions of degrees, becomes plasma, and radiates x-rays. Energetic collisions eject some of the particles from the disk. Often, these charged particles spiral away in tight orbits in the black hole's magnetic field. Two *jets* of high-speed particles can shoot out from the black hole's north and south magnetic poles. As much as ¼ of the material approaching a black hole may be ejected in jets, with the balance ultimately falling through the event horizon and into the singularity. If a steady supply of material falls toward the black hole, a steady stream of particles will shoot out in jets, as illustrated on the right side of Color Plate 15. But if a large mass falls in all at once, the jets become brief bursts. The left side of Color Plate 15 is an actual image taken by NASA's Chandra x-ray space telescope. It shows a central black hole and two bursts of matter jetting away. One jet is coming our way, which makes it appear much brighter. The central bright spot in this image is not the black hole itself (that is black), it is x-ray emission from the black hole's accretion disk. From this distance, we cannot resolve the disk's structure; it looks like a single dot.

Thus surprisingly, not everything that approaches the event horizon falls into the black hole; some material is ejected before being consumed. See Color Plate 16 for an example closer to home.

The right side of Color Plate 17 shows a closeup view of the accretion disk surrounding the central black hole in galaxy NGC 4261. The galaxy is 100 million light-years away, and has a black hole at its center with a mass of over 1 billion *Msun*. On the left is a wide-angle view of the entire galaxy with two jets shooting out above and below. Each jet is nearly 100,000 light-years long—long enough to span our entire galaxy.

The most stunning black hole image may be of Cygnus A in Color Plate 18. The central bright spot is the accretion disk around a black hole. On either side are two enormous jets, streams of particles moving at an estimated ⅓ the speed of light. The jets blast through gas clouds like immense flame-throwers. From end to end, the mayhem spans 500,000 light-years, five times the size of our galaxy. All this is created by an object smaller than a trillionth of a trillionth of the size of an atom.

BLACK HOLES AT GALAXY CENTERS

Astronomers now believe that nearly all major galaxies have super-massive black holes at their centers. Not surprisingly, the most massive objects in a galaxy fall to its center, which is the bottom of the galaxy's gravitational potential.

Color Plate 19 contains two images of a collision of two galaxies. The left image is taken in visible light by the Hubble Space Telescope. On the right, the same galaxies are imaged in x-rays by Chandra. In the Chandra image, the two bright white spots are the accretion disks of the super-massive black holes that occupied the centers of each galaxy before the collision. These black holes are now spiraling toward one another. In about 200 million years, future astronomers on Earth may see them merge and form a single, even more massive black hole. These galaxies are located 400 million light-years from Earth. Therefore, what we see now is what the galaxies were doing 400 million years ago. That is how long it takes their light to reach us. The merger we expect in 200 million years has already occurred there, 200 million years ago. Hope it was (will be) a good show.

SUPER-MASSIVE BLACK HOLE IN OUR GALAXY

Sagittarius A* is a massive black hole at the center of our own galaxy, the Milky Way. It's named in part for the constellation nearest its direction in the sky, and is known to its friends as Sag A*. Compared with other galactic monsters, Sag A* has a modest mass of 4 million *Msun*. Color Plate 20 is a *radio telescope* image of the central two light-years of our galaxy. The red oval near the center is radiation emitted by material swirling around the black hole.

The center of our galaxy is like a shooting gallery. The density of stars there is 100 million times more than in our neighborhood. Some stars are orbiting Sag A* at velocities of several million mph. At such high speeds and densities, accidents are inevitable. Near the galactic center, stars collide or pull one another out of orbit much more frequently than here in the galactic suburbs. Occasionally, a star comes too close to Sag A*. It is then shredded by tidal forces and eventually consumed. Its demise is announced by a blast of radiation.

We learned that stars do not last forever. What about black holes?

Very, very close, but not quite forever.

HAWKING SAYS BLACK HOLES EVAPORATE

The famous British physicist Stephen Hawking discovered that all black holes eventually evaporate via *Hawking radiation*. But don't hold your breath. As discussed earlier, Quantum Mechanics allows very small deviations dE from perfect energy conservation for very brief times dt provided $dt \times dE < h/2\pi$. Enough energy can be "borrowed", according to Heisenberg, to allow spontaneous creation of virtual particle-antiparticle pairs anywhere, anytime. Two photons can be spontaneously created just outside a black hole's event horizon and one can escape to safety, as shown in figure 26.2. The escaping photon's energy must be greater than zero to be a real particle. Thus its partner's energy must be negative because eventually energy must be conserved. No one has ever seen a real particle with negative energy, nor does anyone expect them to exist, however, negative energy is possible for virtual particles.

As the real photon flies away from the black hole, its negative-energy virtual partner disappears through the event horizon into the singularity. Negative energy is equivalent to negative mass ($E=mc^2$), hence the black hole's mass decreases. This is the process that Hawking conceived; the real photons that escape are called Hawking radiation. The energies and intensities are so low that this radiation has never been observed, and probably never will be. However, Hawking's arguments are so cogent that almost all physicists believe this really happens. If a black hole can no longer consume normal gas and stars, it stops growing, and eventually Hawking radiation reduces its mass to zero—the black hole evaporates.

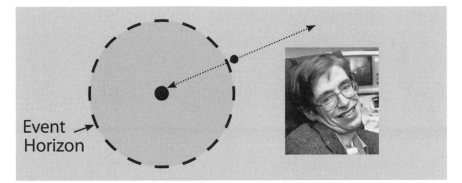

Figure 26.2. Stephen Hawking (1942–) discovered black holes can evaporate by emitting Hawking radiation. Virtual particles will appear near the event horizon (smaller black dot); one particle may escape with positive energy while its negative-energy partner enters the black hole and reduces its mass. The time required for a massive black hole to evaporate is far beyond "astronomical."

Massive black holes emit an infinitesimal amount of Hawking radiation. A black hole with a mass of 2 *Msun* takes 100 billion years to emit enough Hawking radiation to lose the mass of a single proton. The time required for complete evaporation increases with the third power of the black hole's mass and its radiation rate decreases with the mass squared. Therefore, a black hole with twice the mass radiates ¼ as much and lasts eight times longer.

As an evaporating black hole's mass decreases, its radiation intensifies. At the end, the black hole disappears in a flash of Hawking radiation— during the last millionth of a second, it outshines our Sun. A black hole with a mass of 1 *Msun* lasts 10^{67} years after it stops feeding. The largest observed black holes, with masses of 10 billion *Msun*, will last 10^{30} times longer, or 10^{97} years. Such times make the word "astronomical" woefully inadequate. The current age of the universe is roughly 10^{10} years. If the universe lasted 10^{10} more years for every one of its 10^{80} electrons, that would still be 10 million times less than the lifetime of the most massive black hole.

27

NASA's Great Observatories

Astronomy and cosmology have advanced immeasurably due to five magnificent NASA space telescopes. NASA uses the term "Great Observatories" to refer specifically to four space telescopes named after famous scientists: Hubble, Compton, Chandra, and Spitzer. I will add a fifth satellite, the Wilkinson Microwave Anisotropy Probe (WMAP), to this impressive list of NASA accomplishments.

The largest telescopes are ground-based; they are simply too enormous to launch into space. The largest of all are *radio telescopes*, including the two behemoths: Arecibo and VLA. The Arecibo Observatory in Puerto Rico, shown in figure 31.1, is a single 1000-foot-diameter dish (featured in a James Bond movie). The VLA (Very Large Array) in New Mexico has 27 separate dishes, each 82 feet in diameter. Since radio waves easily penetrate our atmosphere (just turn on a radio to hear proof), there is little need for radio telescopes in space. The largest optical telescopes are the twin 31-foot-diameter Keck telescopes operated by Caltech atop Mauna Kea on the Big Island of Hawaii. For many applications, nothing beats the light-collecting power of a big mirror. Because of this, cutting-edge research will always continue on ground-based telescopes, and bigger and better telescopes are always being planned.

But for many imaging applications, our atmosphere is an obstacle. Absorption, turbulence, and interference distort images, degrade

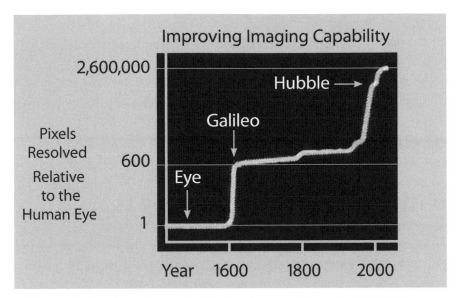

Figure 27.1. How imaging resolution has improved over the last 400 years. The Hubble Space Telescope is the greatest advance since Galileo aimed the first telescope toward the night sky. The vertical axis is the number of pixels resolved relative to the human eye. Adapted from a NASA chart.

resolution, and reduce sensitivity. Our atmosphere protects us from radiation and is convenient for breathing, but it also prevents ground-based observers from seeing celestial sources such as gamma ray emitters. The best place to observe the cosmos is often from space itself. In the 1980s, NASA initiated its Great Observatory program to launch an array of satellites with unprecedented resolution over the broadest range of wavelengths. This program achieved the greatest advance in astronomy and cosmology since the discoveries of Galileo Galilei in the 1600s and Edwin Hubble in the 1920s, which we will discuss in chapter 33.

Figure 27.1, based on a NASA chart, shows the advance of imaging resolution during the last 400 years. Before Galileo, astronomy was done solely with human eyes that can resolve angles of about 1/50TH of a degree. This means if two light bulbs are placed a foot apart, we would see two separate lights only if we are within 3000 feet. Farther away, the angle between the bulbs would be too small and we would see only a single, fuzzy spot of light. The Moon covers an angle of ½ degree in the sky.

As this is 25 times our visual resolution, our eyes see the Moon with $25 \times 25 = 625$ picture elements (*pixels*). If your TV's resolution was that poor, you would turn it off and read a book.

When Galileo became the first to see the heavens through a telescope, he was able to resolve much smaller angles and see 600 times more detail. He realized the Moon is a world of its own, not just a disk in the sky. He saw shadows on the Moon cast by its own mountains. Galileo was the first to see the moons of Jupiter and saw they are orbiting Jupiter and not Earth. This was the first indisputable proof that the geocentric theory of the cosmos was wrong—everything does not, in fact, revolve around Earth.

Over the next 400 years, steady progress was made with ever larger, Earth-based telescopes. But as figure 27.1 shows, the Hubble Space Telescope made all these improvements seem modest indeed. That is, after it was repaired. As launched in 1990, Hubble's main mirror had been ground improperly and its vision was blurry. During a special NASA Space Shuttle mission in 1993, astronauts installed a correcting lens made by the Jet Propulsion Laboratory (JPL), which is operated for NASA by Caltech. It worked admirably well. With continuing improvements, Hubble's resolution is about $1/70,000^{TH}$ of a degree, providing several million times more detail than can be seen by eye.

Color Plate 21 is an image taken by the Hubble Space Telescope commemorating its $100,000^{TH}$ orbit.

But one telescope is not enough. Every telescope sees a certain range of wavelengths; none detects every wavelength. Hubble observes from the near-ultraviolet, through the visible, to the near-infrared. The Compton Gamma Ray Observatory (CGRO) was launched in 1991 to observe the very high-energy end of the spectrum, as its name implies. Unfortunately, CGRO had a gyroscope failure and was "de-orbited" in 2000. NASA replaced it in 2008 with the Fermi Gamma-ray Space Telescope (FGST) that has far greater capabilities than CGRO (see figure 6.3 for an image of Enrico Fermi). The Chandra X-Ray Observatory, launched in 1999, observes primarily low-energy x-rays, and the Spitzer Space Telescope, launched in 2003, observes the infrared portion of the spectrum. WMAP was launched in 2001 to observe the *Cosmic Microwave Background*

(CMB) radiation and has achieved spectacular success. Thus NASA's space telescopes cover the spectrum of light from gamma rays down through microwaves. Coupled with ground-based radio telescopes, we can now observe essentially all wavelengths.

For the first time, we can see every cosmic light source. Why is this important? Different wavelengths probe different cosmic phenomena. Black hole jets emit different light than do accretion disks. The *Big Bang* emitted still different light. We cannot see the center of galaxies at some wavelengths, but we can at others. To get the whole picture, we need to observe all the "colors" of light. Color Plate 22 is a Crab combo—a composite image of the Crab Nebula taken by NASA's Hubble, Chandra, and Spitzer telescopes. The Spitzer image in red probes the nebula's outermost layers, the Hubble image in yellow probes the middle range, and the Chandra x-ray image in white probes the nebula's core, where a neutron star resides. It's not shown in this image, but radio telescopes have also added another essential dimension to our understanding of the Crab Nebula. Astronomers want all the pieces of the puzzle.

We live in the Golden Age of Astronomy
due in large part to the tremendous store of knowledge
collected from marvelous NASA space telescopes.

*"The most incomprehensible thing about the universe is that it **is** comprehensible."*

— *Albert Einstein*

PART 3

The Universe

Our Universe

Redshift

Expansion

The Dark Side

The Big Bang

What Came Before?

28

How Much? How Large? How Old?

We all know our universe is immense. But just how immense is it?

First, let's consider how much it contains. We'll start with something close to home: Color Plate 23 shows an image of our Sun. The small white dot in the upper right was added to show how small Earth would be if it were in this image—can you see it? Our entire world is barely perceptible next to a fairly average star. The spectacular prominence flaring out from the Sun is 40 times larger than Earth. Fortunately, Earth is never this close to the Sun. If it were, we would all need SPF 999,999 sunscreen. The Sun is our nearest star, and it sustains our lives. But it is only one of so many, many stars.

GALAXIES AND THEIR COLLISIONS

Looking out into the cosmos, we see beautiful galaxies—giant assemblies of stars like our Sun. Figure 28.1 shows an image of Andromeda, our largest neighboring galaxy. Andromeda is a classic *spiral galaxy* and is somewhat larger and more massive than our own Milky Way. It is also the largest and most distant object visible to the naked eye. Andromeda is estimated to have a mass of 700 billion *Msun* and to contain a trillion

Figure 28.1. The Andromeda Galaxy is the largest and most distant object observable with the naked eye. It is our largest neighboring galaxy and contains a trillion stars. Image by John Lanoue.

stars. It is 2½ million light-years away and is moving toward us at over 250,000 mph. Andromeda and our galaxy, the Milky Way, will collide in about 3 billion years, and will merge over the following few billion years to form a single, very massive galaxy.

While stars are generally very far apart compared to their size, the distance between galaxies is often only a few times their diameters. Thus galactic collisions and mergers are common. That's how big galaxies get big. The Milky Way may have absorbed several dozen small galaxies to reach its current status—the second-largest behemoth in the Local Group of over 40 galaxies. As with stars, there are far more small galaxies than large ones.

When galaxies collide, they can pass completely through one another without any of their stars colliding. This is because there is so much empty space between stars. The nearest star to our Sun is over 4 light-years away—30 million solar diameters. If people were spaced in the same proportion, there would only be one human on Earth. If a star from Andromeda were to pass through our neighborhood, the probability of

Figure 28.2. Two galaxies in galaxy cluster Abell S0740. The cluster is dominated by the immense elliptical galaxy on the left. Images by J. Blakeslee,
WSU, NASA Hubble, ESA.

the Sun being hit is about the same as the probability of you being hit
by the next meteor to land on Earth. Of course, there would not be just
one Andromeda star passing through; there could be billions of them.
Still, the odds of the Sun being hit are very low.

While star-on-star collisions are unlikely, the gravitational shock of
immense, fast-moving galaxies will wreak havoc on the peaceful, stable
orbits of stars in both galaxies. Vast numbers of stars will be thrown out
into the great beyond, while other stars will be flung into super-massive
black holes. There will be enormous *starbursts*, vast numbers of new
stars created by the gravitational shock of the galactic collision, as well
as a tremendous increase in extremely violent events such as supernovae and radiation from accretion disks. An example of this is shown in
Color Plate 24. Notice the large blue areas in the colliding Mice Galaxies.
Blue light comes from very massive, young stars that shine brilliantly,
are short-lived, and die violently, as discussed in chapter 18.

Color Plate 25 shows the stunning aftermath of another galactic close encounter. Galaxies M81 and M82 are 150,000 light-years apart and 12 million light-years from us. They are locked in a gravitational embrace that periodically brings them close together, setting off spectacular fireworks as seen in M82, the Cigar Galaxy. In addition to starbursts, the Cigar Galaxy displays light from polycyclic aromatic hydrocarbons (*PAHs*). *PAHs* are possible building blocks for the most basic forms of life, although they are considered carcinogenic for humans.

When Andromeda eventually arrives, which side of our galaxy becomes ground zero will make a great deal of difference to our solar system (and to us, if we are still here). We cannot yet predict where the impact will occur—let's hope we're in the cheap seats, far from the action. But all that is 3 billion years from now. Let's get back to today.

THE DISTANT UNIVERSE

I would love to show you an image of our Milky Way comparable to that of Andromeda, but we would have to move our telescopes a million, million, million miles out into space to get the same panoramic view. Instead, Color Plate 26 shows a remarkable NASA/Caltech reconstruction of what we think our galaxy looks like from the outside based on what we observe from the inside. The Milky Way is a **barred spiral galaxy**, with a central bar connected to two major spiral arms, each with several spurs. It is 100,000 light-years across and Earth is perfectly situated 27,000 light-years from its center. The Milky Way has a mass of 600 billion *Msun*, somewhat less than Andromeda.

Looking farther out into the cosmos, we see that galaxies come in a wide variety of sizes and shapes, as seen in Color Plate 27 and figure 28.2. Astronomers are fond of saying that galaxies are like people, they all seem normal, until you get to know them. The largest galaxies are often *elliptical galaxies*, such as the galaxy on the left side of figure 28.2. These are oval-shaped, lacking disks and spiral arms. Ellipticals are thought to be older galaxies that formed from the merger of numerous smaller galaxies.

Many of these very distant celestial objects were first identified by

Figure 28.3. The Sombrero Galaxy is 28 million light-years away and has a mass of 800 billion suns. Image by NASA Hubble.

American astronomer George Ogden Abell, who began his career as a tour guide at Griffith Observatory. He did both his undergraduate and graduate studies at Caltech, and ultimately became chairman of the Astronomy Department at UCLA. Abell is best known for his catalog of galaxy clusters, first published in 1958.

Figure 28.3 is an image of the Sombrero Galaxy that has a mass of 800 billion *Msun*, even more than Andromeda. It has a black hole at its center with a mass of 1 billion *Msun*. The Sombrero is so far away that its light takes 28 million years to reach us. Thus this is really not an image of the Sombrero as it is today. This is what the Sombrero looked like when the light arriving at our telescopes today began its journey 28 million years ago—millions of years before human beings existed.

Looking out farther, we find that galaxies are often in clusters, some containing thousands of galaxies. Figure 28.4 is an image of a cluster 400 million light-years away. On the left, we see the aftermath of a close encounter between two galaxies. A long filament of stars has been torn loose, as if it were cotton candy, and is now stretched between the galaxies. The light we see today began its journey before dinosaurs walked the Earth.

Figure 28.4. Galaxy cluster Abell 1185 is 400 million light-years away. On the left, a stream of stars has been torn from two galaxies that collided. Image by J.C. Cuillandre, CFHT.

Figure 28.5 shows an even more distant cluster; its light has taken 3.2 billion years to reach Earth. The cluster is over 2 million light-years wide and displays some impressive Einstein rings (which were discussed in chapter 22). The galaxy in the upper left is moving through the cluster at an estimated speed of 2 million mph, leaving behind a trail of stars that are being torn loose by the immense gas cloud in which the cluster is embedded. That gas cloud is visible only with infrared imaging.

Color Plate 28 shows the spectacular Hubble Ultra Deep Field (HUDF) image in which we can see almost to the very edge of our universe and almost to the very beginning of time. Light from some of the galaxies seen here has taken 13 billion years to reach us. That journey began billions of years before the Sun and the Earth even existed. The area covered by the HUDF image is less than a tenth of a millionth of the entire sky. But in this very small area, there are over 10,000 galaxies that had never been seen before.

Figure 28.5. Galaxy Cluster Abell 2667 is 3.2 billion light-years away and 2.4 million light-years across. A large Einstein ring is seen right of center. Image by NASA Hubble, J. Kneib, ESA.

The HUDF image is literally a shot in the dark. In 2003, the Hubble team decided to take a gamble, hoping to see farther than anyone ever had before. They chose to invest precious telescope time and risk coming up empty-handed. Imagine, for a moment, that galaxies are trees. We are in a tree in the middle of a great forest. It's hard to see distant trees because so many closer trees block our view. To see very distant trees, we have to find a gap between the nearby ones.

Most "Kodak moments" are captured in a 1/60TH of a second or less. The Hubble team aimed the world's finest telescope at a seemingly empty spot in the sky and opened the shutter for a total of one million seconds (a 12-day exposure taken in small increments over a one month period). Aiming at empty space let them avoid the glare of nearer and brighter objects, but they didn't know what, if anything, they would see. After using all that valuable telescope time, they could have ended up with a blank image; that part of space could have been as empty as it first appeared. But it was not.

The gamble paid off. Hubble saw farther than ever before, and captured an image that casts new light on our understanding of the universe. What a magnificent achievement!

COUNTING STARS

With the HUDF image and others, we are confident that our universe contains over 100 billion galaxies, each with an average of 100 billion stars. Astronomers fondly say:

> There are more stars in the heavens than grains of sand on all the beaches of all the oceans of the world.

I estimate there are at least 1000 times more stars in our universe ($>10^{22}$) than grains of sand on all the world's beaches (10^{19}). That is how immense our universe is in terms of what it contains.

A RULER TO MEASURE THE UNIVERSE

Now, let's discover how immense our universe is in terms of physical size. What is the diameter of our universe? To answer this question, we'll create a ruler to measure the universe. Let's begin close to home, with this heavenly body.

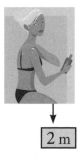

2 m

We'll round off numbers for convenience and say an adult person is 2 meters tall. (Sorry to switch units, but the numbers work out better this way.) Our ruler starts with only a single digit.

9,002 m

Mount Everest is about 4500 times taller than a person. Going to the top of Mount Everest adds three more digits to our ruler.

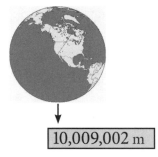

10,009,002 m

Earth's diameter is over 1000 times larger than the height of Mount Everest. Extending our ruler to span Earth adds four more digits.

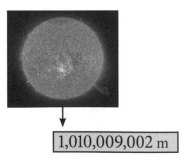

1,010,009,002 m

The Sun's diameter is 100 times that of Earth, adding two more digits to our ruler.

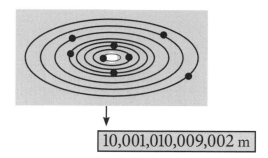

10,001,010,009,002 m

The diameter of our solar system is about 10,000 times larger than the Sun. Getting to the edge of our solar system adds four digits.

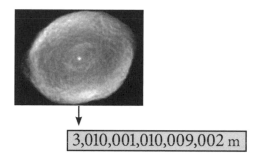

3,010,001,010,009,002 m

The Spirograph Nebula (what our Sun might look like in 5 billion years) is 300 times larger than our solar system: add two digits.

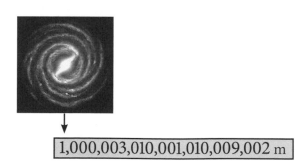

1,000,003,010,001,010,009,002 m

The diameter of our galaxy is 300,000 times larger than the Spirograph Nebula, moving us left six more digits.

$\mathcal{U}$niverse

↓

| 800,001,000,003,010,001,010,009,002 m |

And finally, the diameter of our observable universe is 800,000 times larger than that of the Milky Way. Add five digits to our ruler. Altogether, our ruler for measuring the immense has 27 digits.

But we're not done yet. In our universe, the immense and the minute are intimately connected. To understand the vast, we must also understand the infinitesimal.

Let's start again at the size of a person and now create a ruler to probe the minute, as seen below. A single cell in the human body is 200,000 times smaller than a person. The smallest atom is 100,000 times smaller than a human cell. A single proton, the nucleus of a hydrogen atom, is 100,000 times smaller yet. And finally, the smallest possible distance we think can exist in our universe, the ***Planck length***, is 50 million, million, million times smaller than a proton. That's a very large gap between a proton and the Planck length: 20 digits. We have found nothing whose size lies in this range. We will shortly discuss why we think there is a smallest possible distance.

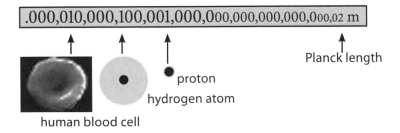

Our ruler for measuring the infinitesimal has 36 digits.

Now, let's stretch our two rulers across the full range of the universe, from the most immense to the most minute.

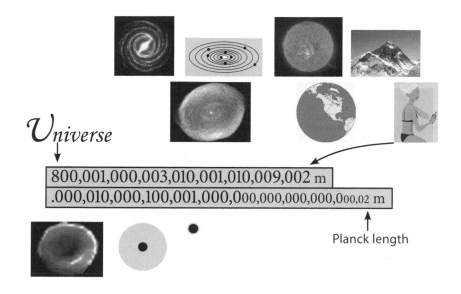

*U*niverse

800,001,000,003,010,001,010,009,002 m

.000,010,000,100,001,000,0₀₀,₀₀₀,₀₀₀,₀₀₀,₀₀₀,₀₂ m

Planck length

Altogether our rulers span 62 digits, 62 factors of ten. This is the distance scale of the universe. It's amazing that the human brain can even begin to fathom such a vast range. On this scale, humans are near the middle of the range. We are larger than the smallest things by 36 digits, and smaller than the largest things by 26 digits.

The smaller digits at the smaller sizes reflect our level of understanding. Throughout the 46 digits at the top, we believe we have excellent physical theories, confirmed by high-precision observations. This is where we are most confident. Across the next 12 digits, we believe our theories are probably still valid, but our current instruments cannot probe such small sizes, so here we are less confident. Finally, over the last 4 digits at the bottom of the range, we are certain that our theories don't work. Here we know nature has secrets we have yet to learn, and that makes this the most exciting area for research, and hopefully, for discovery.

Now, why is there a smallest distance? Combining Quantum Mechanics and General Relativity provides the answer. Quantum Mechanics says that ***virtual particles*** are spontaneously created always and everywhere. The larger the virtual particles' mass, the briefer their existence, and the shorter the distance they can travel. General Relativity says that if a large enough mass is contained within a small enough volume it forms a black hole. Put these two ideas together: if a large mass is spontaneously produced, it is restricted to a very small volume, and makes a black hole. The event horizon of this black hole has an irreducible size. Doing the math, we find the smallest distance, the Planck length, is 1.6×10^{-35} meters. Nothing can be smaller. To be smaller, an object would have to have a shorter wavelength. But it would then have more energy in a smaller volume, and would immediately become a black hole with an event horizon at least as large as the Planck length. Don't worry about space being full of black holes. Black holes with such small masses evaporate in less than 10^{-40} seconds by Hawking radiation. Their lifetimes are infinitesimal, even compared to how long it takes light to move across the width of a proton—they don't last long enough for anything to fall in.

TIME SCALE OF THE UNIVERSE

$\mathcal{U}niverse$

400,000,003,000,000,000.000,000,000,000,
000,000,000,000,000,000,000,000,000,000,000,05 sec

This is a ruler for the time scale of the universe that also spans 62 digits from the very smallest possible time, the **Planck time**, to the very longest time, the age of the universe itself. The smallest possible time interval is 5×10^{-44} seconds, how long it takes to move the smallest distance (the Planck length) at the highest speed (the speed of light). We believe the universe came into existence during one Planck time. On this cosmic time scale, humans are near the top of the range. A human lifetime is a relatively long-lasting phenomenon. Our lives span 53 more digits than the smallest time and only 8 digits less than the age of the universe.

TEMPERATURE SCALE OF THE UNIVERSE

100,000,000,000,000,000,000,000,000,
000,000.000,000,000,000,000,06 K

The temperature scale of the universe spans 50 digits. The highest temperature is 100 million, million, million, million, million degrees Kelvin, the temperature of everything at the instant the universe came into existence. The lowest temperature is 60 millionths of a millionth of a millionth of a degree Kelvin, the equivalent temperature of Hawking radiation on the event horizon of the most massive black hole. Temperature can be related to an equivalent energy, and Einstein showed energy is equivalent to mass. Thus the highest temperature can be related to a mass. By now, you may be able to guess what this is called: the *Planck mass*. It's about ⅓ of a millionth of an ounce (22 micrograms). Of all Planck's bizarre numbers, the Planck mass is the closest to the human scale.

At the instant it came into existence, the total energy of our universe may have been equivalent to only one Planck mass, less than a millionth of an ounce. How is that possible? With such a vast number of stars, each with an immense mass, this seems ridiculous! Indeed, there is tremendous mass-energy in the universe, but there is also an equally tremendous negative, gravitational potential energy. Every star is in the gravitational field of every other star and, therefore, has a negative potential energy—it would take energy to "lift" each star out of the gravitational field of everything else. Doing the math, we find that:

The total energy of our universe is zero.

(Or at least very close to zero.)

This means a universe can be created from almost nothing!

29

What Is Our Universe?

By now you may be asking, what exactly is "*our universe*"? The definition physicists use is: our universe is everything we could possibly observe by any conceivable means, or be influenced by in any conceivable way. It's a practical definition. Science deals with observations and testable theories; it really cannot deal with things that cannot possibly be observed and that cannot possibly have any effect on us. For now, we will stick to the knowable. In chapter 40, we will venture into some speculations about what, if anything, might be beyond and might have come before our universe.

In figure 29.1, we see Earth at the center of a circle. The circle should really be a three-dimensional ball (but the printer had a problem with that). The radius of the circle is how far light could have traveled since the universe began. Nothing can travel through space faster than the speed of light: no particle, no interaction, and no influence. Light is extremely fast—671 million mph—and the universe is very old—13.7 billion years old. But as big as those numbers are, the distance light could have traveled since the beginning of time is still limited; it is at most 13.7 billion light-years, which is the radius of the circle in figure 29.1 (it's obviously not drawn to scale).

We can conceivably observe anything within the circle that defines our universe; we might need a telescope as large as the solar system, but

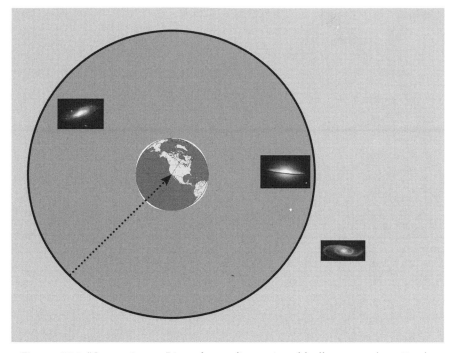

Figure 29.1. "Our universe" is a three-dimensional ball centered on Earth. Its radius is how far light can have traveled since our universe began. The galaxy on the right is not in our universe—it is too far away for its light to have reached us by now.

at least it's possible. However, we cannot possibly see the galaxy on the far right; its light simply has not had enough time to get here. By our definition, that galaxy is not in our universe, at least not yet.

As figure 29.2 shows, other observers' universes may vary. Astronomers in another galaxy would define a different circle as their universe; it would be centered on them. If we can see their location, then they can see our location and we would be in each other's universes. The galaxy on the far right would be in their universe even though it is not in ours. Similarly, there would be galaxies in our universe that are too far away for them to have seen by now.

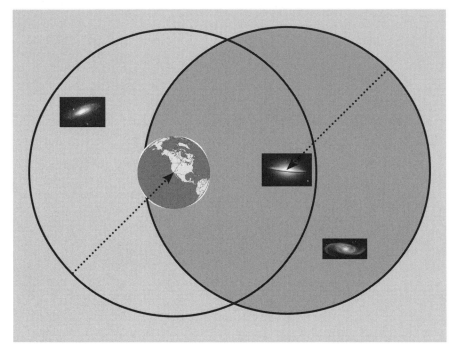

Figure 29.2. Astronomers in another galaxy would say their universe is centered on their location and contains a different set of galaxies.

WHAT, IF ANYTHING, LIES BEYOND?

Is our universe all there is?

No cosmologist believes that. But because we cannot see any farther, we cannot be sure. Probably there is much, much more beyond the edge of our universe. Perhaps what lies beyond is as vast compared to our universe as our universe is compared to an atom.

Is the stuff beyond our universe just like the stuff within our universe?

Again, we cannot know for sure: it probably is. But it's possible that there could be something very different outside our universe. Maybe even something sinister, as in figure 29.3.

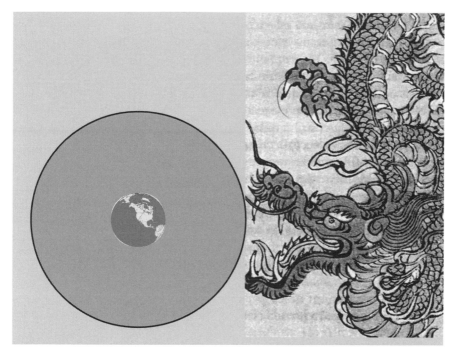

Figure 29.3. Could there be something very different beyond our universe? Maybe even something sinister? No worries, mate.

But you can rest easy. As we will discuss later, space is expanding so rapidly that any distant monster is not going to get out of the closet.

In the next few chapters, we will discuss the most important facts that we know about our universe, and examine the actual, primary scientific evidence for these facts. This will allow you to judge for yourself the strength of this evidence.

In chapter 39, we will explore science's best theory of our universe—the Inflationary Big Bang theory—which puts all these facts together into a coherent story.

30

Telescopes Are Time Machines

The farther we look out into space, the farther we are also looking back in time.

As hinted earlier, telescopes are a type of time machine. They don't take our bodies back to the good old days (whenever that was), but they do take our eyes there and let us see the cosmos as it actually was in the distant past.

> Telescopes show us the past as it *is* actually happening.

Color Plate 19 shows an image of two galaxies 400 million light-years away. This means light from these galaxies takes 400 million years to reach Earth. We do not see the galaxies as they are today; we see them as they were 400 million years ago—in the throes of a cataclysmic collision. Now, all the excitement is probably over, over there. The galaxies probably merged long ago and became one big happy galaxy. But we are not over there. Here, where we are, this collision is happening now. As astronomers continue observing this spectacle, they will see the collision and subsequent merger proceeding as it really happens(ed). But what we see here happens 400 million years after it happened there.

Figure 30.1. Magnified portions of NASA's Hubble Ultra Deep Field image show the formation of some of the earliest galaxies nearly 13 billion years ago.

The ability to see the cosmic past actually unfolding before our eyes is a tremendous boon to science. Astronomers don't have to wonder when or how galaxies formed. They can directly see what, when, and how everything happened because live images have been preserved in the time capsule of light traveling across the vastness of space. We know small assemblies of stars began forming galaxies over 13 billion years ago. We know these gradually grew through mergers. We know that the process of galaxy formation was initially extremely violent and accompanied by the release of a tremendous amount of energy. We know the universe gradually matured and became more placid.

We know all this, not through a series of clever logical deductions, but because, in our telescopes, we see the cosmos in all these different stages of development.

Figure 30.1 contains magnified portions of the Hubble Ultra Deep Field image shown in Color Plate 28. In these magnified views, we see some of the first galaxies in the process of forming 13 billion years ago. We know when this happened and what the first galaxies were like because we actually see them in action.

Figure 22.1 shows an image of the Einstein Cross. We see a super-luminous quasar gravitationally lensed by an intervening galaxy. We believe quasars no longer exist, that all the quasars in the universe stopped "quasing" long ago. "Quasing" is something galaxies do in their awkward adolescence, when black holes in "teenage" galaxies voraciously consume nearby gas and radiate energy profusely. They devour everything in sight and make a huge commotion that blasts out in all directions. (Does this remind you of anything?)

After billions of years, galaxies mature and become more sedate.

Most of their stars and gas clouds have either found stable orbits or have already been consumed. Their black holes' feeding frenzies abate and super-luminous emissions wane. Fortunately, telescopes allow us to observe these spectacular monsters that no longer exist, while keeping a safe distance.

Similarly, if astronomers in distant galaxies are observing us, they would see our past, not our present.

Imagine all the questions we could answer with certainty if we could see real images of Earth's past. How did the Moon form? What were the key developments in human evolution? How did the Egyptians build the pyramids? Is the story of Noah's Ark based on an actual flood? What happened to the Mayan civilization? Did Tony Soprano get whacked?

Imagine if paleontologists could use telescopes to view dinosaurs on Earth. They would actually see them as they really lived, as figure 30.2 purports. They would directly see dinosaurs' behaviors, their gaits, and their true colors. They wouldn't have to wonder whether dinosaurs cared for their young, or just ate them.

Maybe dinosaurs were smarter than we think.

Figure 30.2. If only a telescope could see what dinosaurs were really like.

Color Plate 1. Color is critical in the photoelectric effect. Einstein explained that because blue light has a higher frequency (and thus greater energy), it can eject electrons (green) from a metal (gray). However, red light can't do this, even if it is much more intense, because its frequency is too low.

Gamma rays	X rays	UV	Visible	IR	Micro- waves	Radio waves

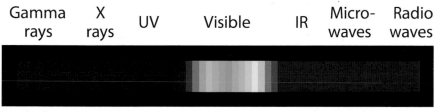

High frequency
Short wavelength

Low frequency
Long wavelength

Color Plate 2. The spectrum of light ranges from gamma rays with the highest energy to radio waves with the lowest energy. Visible light lies in a narrow band between ultraviolet (UV) and infrared (IR).

H

Hg

Ne

Color Plate 3. The spectra of hydrogen (H), mercury (Hg), and neon (Ne). Atoms of each element emit and absorb only a specific set of frequencies of light, providing a unique fingerprint visible across the entire cosmos.

Color Plate 4. M51, the Whirlpool Galaxy, lies in front of a smaller companion galaxy NGC 5195. Image by Beckwith, NASA Hubble, ESA, AURA.

Color Plate 5. The Eagle Nebula is in green at upper left with wings spread and talons in front. Its wings span 200 trillion miles. Image by NASA Spitzer.

Color Plate 6. Above images by NASA Hubble show clusters of new stars. The stars in NGC 602 (left) are less than 5 million years old, while the stars in NGC 3603 (right) are less than 2 million years old.

Color Plate 7. The Pillars of Creation are up to 6 trillion miles long. Inset is a magnified view of emerging new stars (pink). Images by NASA Hubble.

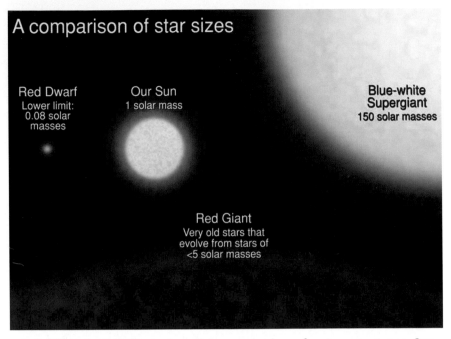

Color Plate 8. NASA illustration of sizes and colors of various star types. Our Sun is now a yellow dwarf but will become a red giant in 5 billion years.

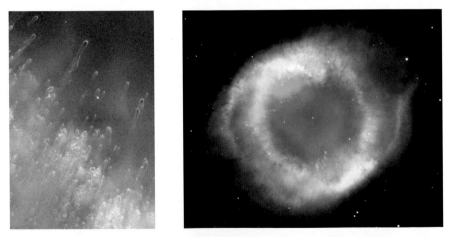

Color Plate 9. Right: In the Helix Nebula, the stellar wind from a new star is clearing the star's neighborhood. Left: Magnified view of nebula's inner rim; only dense objects survive the wind's erosion. Images by NASA Hubble.

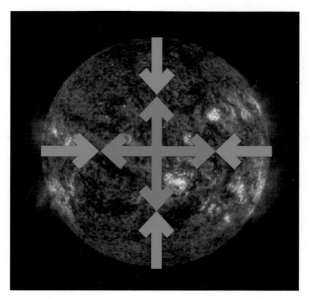

Color Plate 10. The force of gravity (blue) is balanced by pressure (green), making stars stable for billions of years. Image of our Sun by NASA SOHO, ESA.

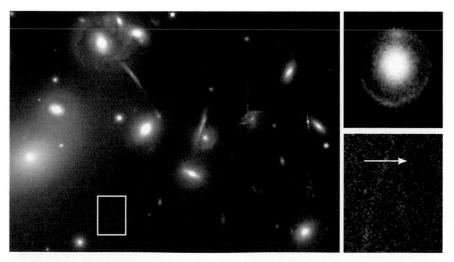

Color Plate 11. Upper right: A nearly perfect Einstein ring (blue). Left: Several partial rings can be seen amid a galaxy cluster. The boxed area is magnified in the lower right; light from the indicated galaxy (red) has traveled toward us for 13 billion years. Image by Richard Ellis, Caltech, NASA Hubble, ESA.

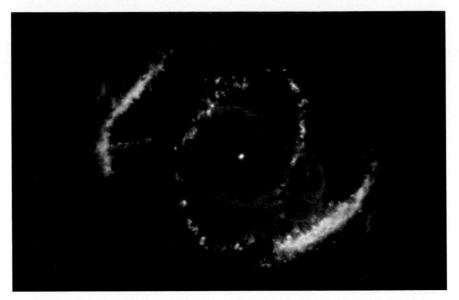

Color Plate 12. The Cat's Eye Nebula, 3000 light-years away, is formed from a star's castoff outer layers that surround its collapsed core, now a white dwarf (central white dot). Image by NASA Hubble.

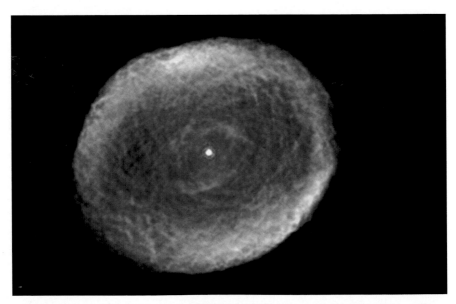

Color Plate 13. The Spirograph Nebula is relatively small, only 2 trillion miles wide. Our Sun may look like this in 5 billion years. Image by NASA Hubble.

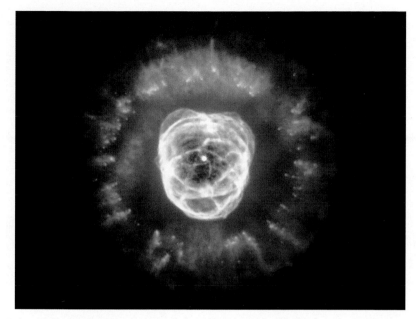

Color Plate 14. The Eskimo Nebula is 5000 light-years away and 8 light-years across. Image by NASA Hubble.

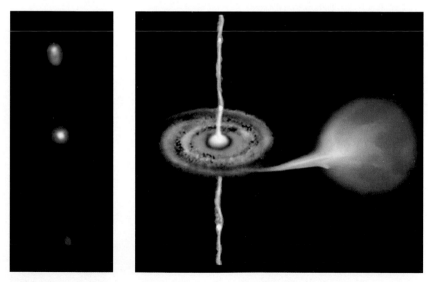

Color Plate 15. Right: NASA illustration shows gas from a companion star (red) falling onto the accretion disk of a black hole. Some material is ejected in jets (white). Left: Actual NASA Chandra x-ray image shows burst jets.

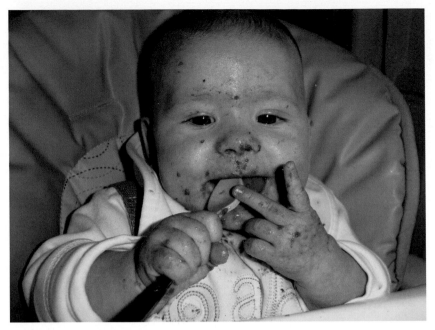

Color Plate 16. Not everything approaching an event horizon is consumed; 25% may be ejected in high-speed jets. The author gratefully acknowledges this demonstration by 6-month-old granddaughter Isabella.

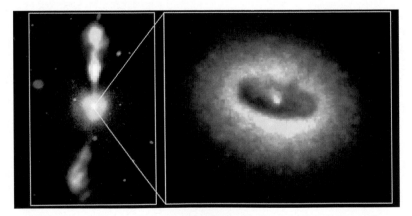

Color Plate 17. Right: An accretion disk surrounds a black hole at the center of galaxy NGC 4261. Left: In a wide-angle view of the same galaxy, each jet is nearly as long as our galaxy, the Milky Way. Image by NASA Hubble.

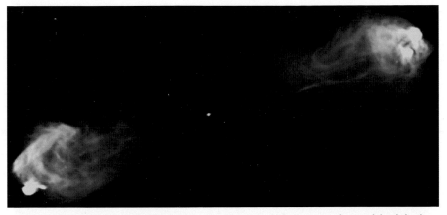

Color Plate 18. The central dot is an accretion disk surrounding a black hole in Cygnus A. Two jets blaze through the interstellar gas like flame-throwers, spanning 500,000 light-years. Image by VLA, NRAO, AUI.

Color Plate 19. Left: NASA Hubble image shows two galaxies colliding 400 million light-years away. Right: An x-ray image of the same galaxies by NASA Chandra shows the two black holes at the galaxies' centers (white dots) that will likely merge to form one black hole in about 200 million years.

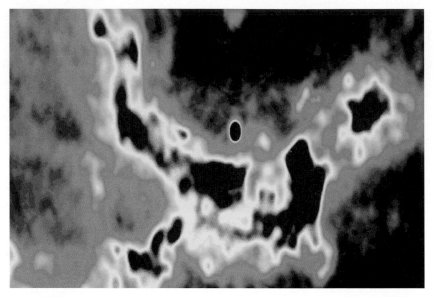

Color Plate 20. A radio telescope image by Northwestern, NRAO, shows the center of our galaxy. Black hole Sagittarius A* lies within the red oval.

Color Plate 21. NGC 2704 is 200,000 light-years away. This image by NASA Hubble celebrates its 100,000^(TH) orbit. Inset is an image taken from the Space Shuttle showing the space telescope named in honor of Edwin Hubble.

Color Plate 22. Composite image of the Crab Nebula by three NASA space telescopes—Spitzer (red), Hubble (yellow), and Chandra (white).

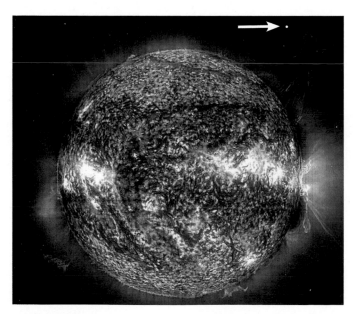

Color Plate 23. An image of our Sun on a bad-hair day by NASA, SOHO, EIT, ESA. The white dot indicated by the arrow shows how large Earth would be on the same scale. Fortunately, we are never this close to the Sun.

Color Plate 24. The Mice Galaxies NGC 4676 after a collision that initiated immense starbursts that are seen in blue. Image by NASA Hubble.

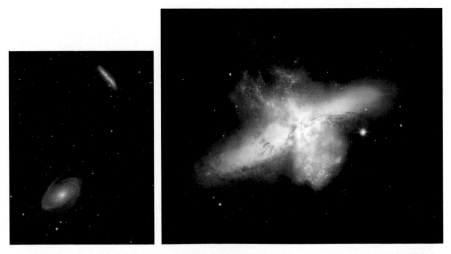

Color Plate 25. Left: Galaxies M82 (above) and M81 are gravitationally bound to one another. Right: A closeup of M82, the Cigar Galaxy, which was disrupted by the last close encounter with M81. Images by NASA Hubble.

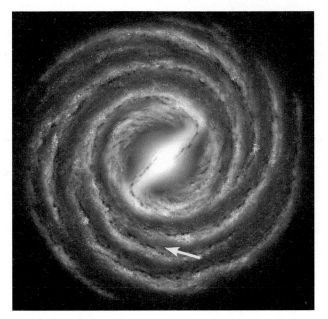

Color Plate 26. NASA/Caltech reconstruction of our Milky Way, a barred spiral galaxy, which is 100,000 light-years across. We are perfectly situated (tip of yellow arrow) 27,000 light-years from the center.

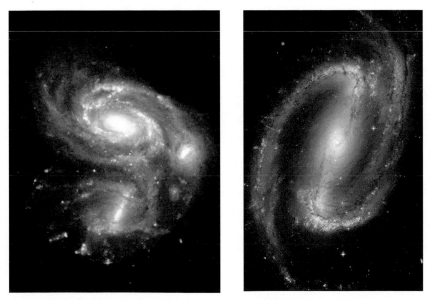

Color Plate 27. Left: colliding galaxies in Arp 272. Right: Galaxy NGC 1300. Images by NASA Hubble, ESA.

Color Plate 28. NASA Hubble Ultra-Deep Field image takes us almost to the edge of our universe and almost back to the beginning of time. There are 10,000 galaxies in this image. Total exposure time was 1 million seconds.

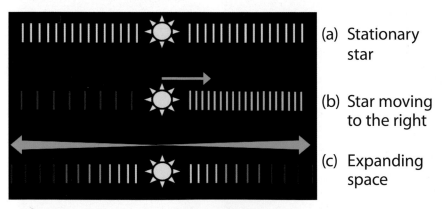

(a) Stationary star

(b) Star moving to the right

(c) Expanding space

Color Plate 29. Color changes in a single wavelength of starlight. In (a), light from a stationary star is yellow. In (b), light is blueshifted in the direction the star is moving and redshifted in the opposite direction. In (c), the expansion of space redshifts the light in proportion to the distance from the star.

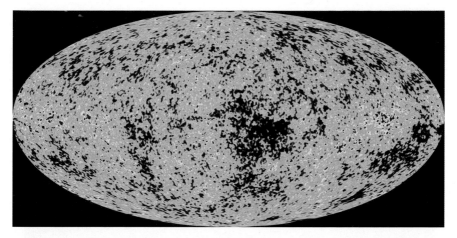

Color Plate 30. NASA WMAP image shows the universe at 1/36,000TH of its current age. The Cosmic Microwave Background (CMB) radiation is extremely uniform—red areas were 1/100,000TH more dense than blue areas.

Color Plate 31. The Bullet Cluster. The galaxy cluster on the right passed through the cluster on the left. While the plasma clouds (red) were slowed by the collision, dark matter haloes (blue) were unaffected. Image by NASA CXC, CfA, STSci, ESO WFI, Magellan, Univ. Arizona, D. Clowe, M. Markevich, et.al.

Color Plate 32. Earth rises over the Moon's horizon. Image was taken from lunar orbit by NASA astronauts on Apollo 8.

Color Plate 33. The Rosette Nebula imaged with three spectral filters identifying specific atoms—oxygen atoms are shown in green. Image by NOAO.

Figure 30.3. Enrico Fermi in the regalia of a hallowed European university professor. Family photograph.

FERMI'S DELICIOUS BOOK

I wish to relate a personal story about Enrico Fermi.

As I mentioned earlier, Fermi was my father's mentor and his thesis advisor at the University of Rome. When I was a baby, our family was invited for dinner at the Fermi's house in Chicago. My parents let me crawl off so the adults could talk in peace. Later, they found me chewing on pages I had ripped from the great scientist's physics books.

Exasperated, but always the gentleman, Fermi said of me: "Well .... at least he has good taste."

It seems my tastes have never changed—I still enjoy devouring a good physics book.

31

It's the Same Everywhere

If the universe were divided into a million equal boxes, the contents of each box would be pretty much the same. Each box would have about the same number of stars and galaxies, and about the same average density, temperature, and pressure. And perhaps most importantly, each box would have the same types of atoms and particles. We observe the same hydrogen atoms emitting the same light in the most distant galaxies as we see here on Earth.

The universe could well have been radically different in different places, but it is not. This is called:

> The Cosmological Principle
>
> On a large scale, the universe is the same everywhere.

The phrase "On a large scale" is a critical part of the Cosmological Principle, as we will discuss shortly.

Figure 31.1 shows an image of the world's largest telescope—the radio telescope in Arecibo, Puerto Rico. Whereas conventional telescopes collect visible light, radio telescopes collect light at radio wave frequencies. (See chapter 14 for a discussion of the spectrum of light.) Shown in figure

Figure 31.1. The world's largest telescope is the 1000-foot-diameter radio telescope in Arecibo, Puerto Rico.

31.2 is a map of the sky made by the Arecibo radio telescope. Since the heavens completely surround us, an image of the entire sky is shaped like the inside surface of a sphere. The figure's oval shape approximates a sphere that has been cut open and stretched flat onto the page. Grid lines have been added for reference. The north celestial pole is at the top of the image. The celestial equator runs horizontally across the map's widest part at the center. Horizontal curved lines are celestial latitude lines. Curved lines of celestial longitude run from the north to the south poles. The map's left edge and right edge correspond to celestial longitudes of –180 and +180 degrees that, like the International Date Line, are in fact the same location on the spherical surface.

The area at the bottom of the image is blank because the telescope is located in the northern hemisphere and cannot see all the way to the south celestial pole. The odd-looking U-shaped, blank region that cuts across the image is where our galaxy, the Milky Way, lies in the sky.

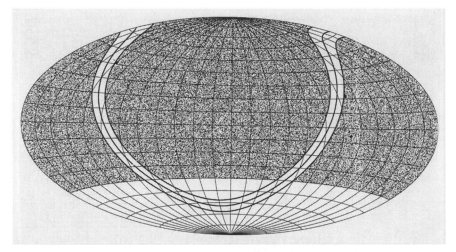

Figure 31.2. A radio telescope image from Arecibo, Puerto Rico, shows the locations of the 125,000 brightest radio galaxies. Bottom white band is too far south for this telescope to see. U-shaped white band is where our galaxy lies in the sky; it was deleted to better view the distant universe. The observed distribution of galaxies seems very uniform, indicating that our universe is homogeneous—the same everywhere.

To examine the distant universe, we need to look past our own galaxy, thus this part of the image is left blank. Other than these two blank areas, figure 31.2 is an image of everything in the heavens that a radio telescope can "see."

The image in figure 31.2 shows the locations of the 125,000 brightest radio galaxies; each galaxy is represented by a single dot. The image shows that the universe is the same everywhere; it is **homogeneous**. There is no apparent center to the universe, no edges, and no clumping.

In chapter 34, we will discuss the cosmic microwave background radiation (CMB) that is shown in Color Plate 30. The CMB also provides very strong evidence for the Cosmological Principle. It shows that, on a large scale, the energy density of the early universe was uniform to within $1/1000^{\text{TH}}$ of 1%.

The Cosmological Principle states that the universe is the same at all locations, but not at all times. Indeed, the universe is quite different today than it was 12 billion years ago, and it will no doubt change again over the next 12 billion years. But as the universe evolves, at each moment, it is the same everywhere on a large scale.

The phrase "on a large scale" is an important qualifier. Clearly, the universe is not the same everywhere on every size scale.

In fact, on every sub-cosmic scale, we see complex structure and vast differences from one place to another. The Milky Way and Andromeda are very different from the nearly empty space lying between these two great galaxies. Within our solar system, nearly 99.9% of all its mass is contained within only 4 trillionths of its volume (in the Sun). In an atom, the concentration of mass is 10,000 times greater yet.

Stars, planets, and humans are a million, million, million, million, million times more dense than outer space.

But on a cosmic distance scale, all these differences average out. On a cosmic scale, even galaxies are tiny. If the universe were divided into a million equal boxes, as mentioned at the start of this chapter, each of these boxes would be immense. The size of each box would be the better part of a billion light-years on a side, and each box would contain 100,000 galaxies. It is on this very large scale that the cosmos is homogeneous.

This large-scale homogeneity is very important to our understanding of cosmology, even as the small-scale differences are vitally important to our existence.

Cosmology begins with Einstein's Theory of General Relativity—that's how we understand gravity, the most important force in the cosmos. Unfortunately, Einstein's equations are very difficult to solve exactly. No one has ever found an exact solution of Einstein's equations for a universe that is highly inhomogeneous. But we do have an exact solution of Einstein's Field Equations for a homogeneous universe, like ours.

The small-scale *inhomogeneity* of the universe enables life.

Its large-scale *homogeneity* allows life to understand it.

32

Redshift / Blueshift

This chapter is about changes in the "colors" of starlight due to motion: *redshifts* and *blueshifts*. Because of these shifts, we can measure the motions of stars, galaxies, and the universe itself. For very remote objects, these color changes are the only way to measure motion, and may be the only way to determine distance. Redshifts and blueshifts are, therefore, of critical importance in astronomy and cosmology.

Consider the stationary star on the left side of figure 32.1. Imagine for simplicity that it emits light of only one wavelength. (What we discover for one wavelength will also apply to all other wavelengths.) The circles in the figure indicate where the light waves are cresting. The largest circle was emitted first, and the distance between circles is equal to one wavelength. (For clarity, the wavelength is greatly exaggerated.) Since the star is stationary, the source of light has not moved and each circle is centered at the same point. Now, examine the star on the right side of the figure, which is moving to the right with a constant velocity. Each circle around the moving star has the same radius as the corresponding circle of the stationary star, but their centers are shifted. Since the star is moving, the starting point of emitted light, the center of each circle, moves progressively to the right as time passes. The most recently emitted light, the smallest circle, is centered near where the star is now. The larger circles are centered where the star was earlier, when that wave was emitted.

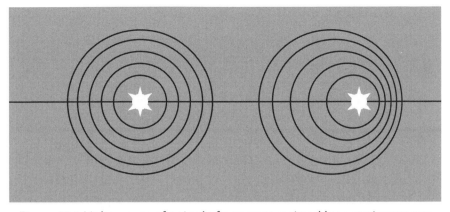

Figure 32.1. Light waves of a single frequency emitted by a stationary star (left) and a moving star (right). Circles are wave crests; each is separated by one wavelength. Horizontal line is shown for reference. Light from the moving star has a shorter wavelength on the right (blueshift) and a longer wavelength on the left (redshift) as compared with the stationary star.

Now, focus on where these circles of wave crests intersect the horizontal line. For the stationary star, the circles cross the horizontal line with even spacing. The wave crest spacing is one wavelength and it is the same on both sides of the star. For the moving star, the wave crests are equally spaced on the right and equally spaced on the left. But the wave crests are spaced farther apart on the left than on the right. This means the light's wavelength is longer on the left than on the right. Equivalently, the moving star's light has a lower frequency on the left than on the right. In the visible part of the spectrum, red has the longest wavelength and blue has a short wavelength. As shown in Color Plate 29, light on the left side of the moving star looks redder than light from the stationary star and light on the right side of the moving star looks bluer. To observers on the left, the star is moving away and its light appears redder; it is *redshifted*. To observers on the right, the star is moving toward them and its light appears bluer: it is ***blueshifted***. The greater the star's velocity, the larger these shifts will be. These color changes allow a star's velocity to be measured. Astronomers can measure these color shifts with extraordinary precision, to parts per billion, determining stars' speeds to within ±1 mph, in some cases.

If a star is moving sideways to our line of sight, the wavelength shift is almost zero. Because of this, only the component of a star's velocity along our line of sight is accurately measurable, but that is usually enough to tell a great deal.

All this explains color shifts that are due to stars moving through space. But what if a star is stationary and space itself is moving? What does it mean for space to move? As we'll see in the next chapter, space can expand or contract: the universe could be getting bigger or smaller. What happens to starlight then?

The lowest image in Color Plate 29 shows what happens if space is expanding. If space expands uniformly, there is no specific point it is expanding away from, rather it expands away from all points. Imagine a balloon that is being inflated. Viewed from any point on its surface, as the balloon stretches, all other points appear to be moving away. Similarly, in the uniformly expanding three-dimensional space of our universe, any star can be seen as the center of expansion, because everything else appears to be moving away from it.

As space expands, its contents are carried along. All the galaxies, stars, particles, energy, and light are carried along by the expansion. The wave crest circles also expand with space, stretching their wavelengths and making their light redder. The larger circles have expanded more because they have been in the expanding space longer. They are more redshifted than the recently emitted light closer to the star. As Color Plate 29 shows, light from a stationary star is progressively more redshifted the farther it travels. If space has doubled [1] since a light wave was emitted, the light's wavelength will also have doubled.

If space is contracting, the opposite happens—light is progressively blueshifted. But as we shall soon learn, redshifts are more common than blueshifts.

Wavelength shifts have become the principal means of describing how far away very distant objects are and how large the universe was in the past.

Recall from chapter 16, each element has a unique light emission and absorption spectrum—a unique fingerprint. If light from a distant galaxy contains the spectrum of hydrogen, but with all the wavelengths doubled, we know this light was emitted when the universe was half its current size. The wavelengths have doubled due to space doubling [1] since the light was emitted. This is like finding an enlarged fingerprint on a balloon. Even enlarged, the fingerprint identifies a unique individual and the amount of enlargement determines how much the balloon has inflated since the fingerprint was made.

The standard terminology for wavelength shifts uses the letter z: $z+1$ is the wavelength expansion factor. Let's consider examples in an expanding universe. If the universe doubled in size during the time light took to reach us from a certain galaxy, then the wavelength of that light would also have doubled. The expansion factor would be 2, hence $z+1=2$, and $z=1$. Astronomers would say that galaxy "is at $z=1$." Light from a nearby source would have a z only slightly more than 0 because the expansion factor would be only slightly more than 1 during the brief time it took light to reach us. At the other extreme (discussed in chapter 34) is light that has $z=1091$; it was emitted when the universe was only $1/1092^{ND}$ of its current size.

Most astronomers and cosmologists use z to describe the distance to very remote light sources because z can be directly measured with great precision, while the distances themselves are not directly measurable. Using z and making certain assumptions, we can estimate how long and how far each ray of light has traveled. If you read about astronomers finding a galaxy "at $z=6$", you can use the table on page 317 to find out how far its light has traveled to reach us and how far back in time we are looking. This table tells us that for $z=6$, we are looking back 12.7 billion years to when the universe was 1 billion years old.

Every ray of light we see from distant objects tells us from its redshift how large and how old the universe was when that light was emitted.

NOTES

[1] "Space doubling" means space becoming twice as large in each of its three dimensions, making its volume eight times larger.

33

Expansion

Einstein published his Theory of General Relativity in 1915, when nearly everyone believed the universe was static—that the universe always was, and always will be, as it is now.

This belief in an unchanging universe, though not supported by any solid evidence, was deeply ingrained in Western thought, just as the belief in a flat Earth had been deeply ingrained several centuries earlier.

From a 21ST century vantage point, the idea of a forever-unchanging universe seems quite bizarre and contrary to our everyday intuition. Things don't simply hang in mid-air without support. Free-moving objects may be going up, if something has launched them, or they may be dropping down, but they don't remain static. In both Newton's theory and Einstein's theory, it isn't possible for freely moving objects to remain stationary in a gravitational field. Their velocities must be continuously changing—they must move.

Since the universe is an assembly of gravitating bodies, it cannot be static either. In a homogeneous universe, galaxies must either be falling toward one another or flying apart, but they cannot remain motionless.

This means that the universe must be either contracting or expanding.

EINSTEIN'S "GREATEST BLUNDER"

Nonetheless, before the 1920s, it was commonly assumed that the universe was static. Soon after publishing his General Theory, Einstein realized it wasn't compatible with a static universe, and he decided to "fix" it. In 1917, he published a revised theory with a new term, the *cosmological constant Λ* (Greek capital letter *lambda*). Einstein's new Field Equations were: $G+Λ=8πT$. The new term was added to provide repulsive gravity—gravity that pushes objects apart rather than pulling them together, as normal gravity always does. Einstein offered no physical explanation for this new term. However, it did allow him to choose a value of $Λ$ that balanced the overall attractive force of gravity, which he hoped would make his theory compatible with a static universe.

It wasn't Einstein's greatest moment. Recall from chapter 16 that Planck's suggestion to avert the "ultraviolet catastrophe" was a solution without an explanation. Einstein's cosmological constant provided neither an explanation nor a solution. With only one number to adjust, the repulsive $Λ$ could, at best, balance gravity only at one moment and only on an average basis. At best, it could provide an unstable equilibrium, as shown on the right side of figure 18.1, and that's not good enough. Like the ball in that figure, even if it is perfectly balanced initially, it cannot remain so for long. Things move: planets orbit stars, stars orbit galaxies, and galaxies orbit clusters. Inevitably, imbalances will develop. Wherever there is even slightly more mass, gravity will overpower $Λ$, and that region will start collapsing, strengthening gravity, and making that region's collapse irreversible. Wherever there is even slightly less mass, $Λ$ will overpower gravity, and that region will expand, weakening gravity, and making that region's expansion irreversible.

In 1922, Alexander Friedman, a Russian cosmologist, mathematician, and meteorologist, published a solution of Einstein's Field Equations proving a static solution was impossible—the universe must be either expanding or contracting.

Einstein's attempt to fashion a static universe was ill-conceived for theoretical reasons—worse yet, it turned out to be wrong.

Figure 33.1. Henrietta Leavitt (1868–1921) discovered a method for measuring vast distances; image courtesy of American Institute of Physics. Edwin Hubble (1889–1953) used Leavitt's method to show the universe is far larger than previously believed, and that it is expanding; image courtesy of NASA.

LEAVITT AND HUBBLE OPEN THE UNIVERSE

In the early 20TH century, astronomers did not know how big the universe was. They thought the Milky Way was all that existed. But in 1922, Edwin Hubble discovered that some of the so-called "nebulae" are really galaxies in their own right and are incredibly far away. Hubble proved Andromeda is not just a gas cloud with "lots" of stars, but is, in fact, a major galaxy comparable to our Milky Way. He proved this by observing *Cepheid variable stars* in Andromeda, and using a relationship discovered by Henrietta Leavitt. Hubble and Leavitt are shown in figure 33.1.

Hubble was raised in Illinois and was a high school track star, placing first in seven events in a single meet. He also set the state high-jump record. Who would have guessed then that he would later leap beyond our galaxy? Leavitt overcame deafness and male chauvinism to become the first famous, female American astronomer. She began her astronomy career as a technical assistant at the Harvard Observatory, assigned to measure the brightness of stars on photographic plates, at a time when women were not allowed to operate telescopes. In particular, she studied *variable stars*—stars that change their brightness in a regular, repetitive fashion. In 1912, Leavitt discovered that variable stars of a certain type, called *Cepheid variables*, obey a special relationship: the period of a Cepheid—how long its brightness takes to cycle up and down—is directly related to its maximum energy output. By measuring how much of a Cepheid's light reaches Earth, and using Leavitt's relationship, astronomers can now determine how far away that star is. This is huge! For the first time, astronomers could measure distances on an intergalactic scale.

Hubble used Leavitt's Cepheid relationship and other measurements to determine the distances to many of the closest galaxies. He went on to compare these distances to the galaxies' redshifts. In 1929, he shocked the world with his discovery that almost all distant galaxies are moving away from us. And the more distant they are, the faster they are moving away. As we will learn, Hubble's discoveries led inexorably to the conclusion that:

> The universe is expanding!

Faced with proof that the universe wasn't static after all, Einstein renounced Λ, calling it his "greatest blunder." As a mere mortal, Einstein was not infallible. In fact, he was wrong on a fair number of occasions. Perhaps his batting average was "only" .500 (a 50% success rate). A .500 average is catastrophic for an airline pilot or a heart surgeon, but it's spectacular for a batter or for a theoretical physicist wrestling with the Big Questions. As Einstein said, "Anyone who has never made a mistake has never tried anything new."

When Einstein got a hit, it was often a grand slam, and when he struck out, it didn't hurt the advance of science—others saw to that. Einstein's "blunders" do not diminish his phenomenal impact on physics. In fact:

> Einstein was World MVP—the World's Most Valuable Physicist

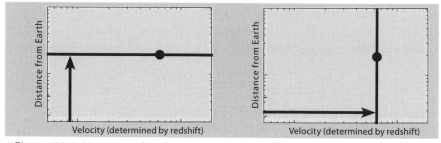

Figure 33.2. Here we plot the distance and velocity of a supernova (black dot). Left: The supernova's distance from Earth determines how far up it is plotted. Right: The supernova's velocity moving away from Earth (measured from its redshift) determines how far to the right it is plotted.

MEASURING THE EXPANSION

The expansion of the universe is one of the most important discoveries of modern science. Let's really dig into this and see how we know that the universe is expanding and what this means.

Today, the most precise measurements of the expansion of the universe are made using *Type Ia supernovae* (recall that Ia is "one-A"). As discussed in chapter 23, all Type Ia supernovae release almost exactly the same amount of light. By measuring how much light reaches Earth, a supernova's distance can be computed, and its velocity can be measured from the redshift of its light. Figure 33.2 illustrates how the distance and velocity of a supernova are plotted. The distance from Earth is plotted on the vertical axis; a supernova's distance determines how far up it appears on the plot. The velocity is plotted on the horizontal axis; a supernova's velocity determines how far to the right it appears on the plot. Since almost all stars are in galaxies, the distance and velocity of a supernova determines the distance and velocity of its host galaxy.

Consider for a moment only those galaxies that happen to be 1 billion light-years from Earth. On our chart, they are all plotted on the same horizontal line because they are all at the same distance. One might think these galaxies would be moving with a wide range of velocities: some toward us, some away, some fast, and some slow, as in the left side of figure 33.3. Why not? What Hubble and subsequent astronomers actually

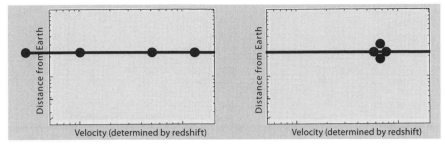

Figure 33.3. Left: how our universe might have been—galaxies might have had a wide range of velocities. Right: how our universe really is—Hubble discovered that all remote galaxies at the same distance are moving away at the same velocity.

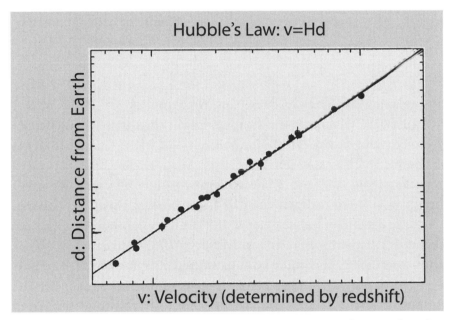

Figure 33.4. Hubble's Law: distant galaxies are moving away from Earth with velocities proportional to their distances. Dots represent Type Ia supernovae from all directions in the sky discovered before the Hubble Space Telescope began its search. The most distant supernova here is at redshift *z=0.1*.

found is that these galaxies are all moving way from us at the same velocity—50 million mph—as on the right side of figure 33.3. Hubble found that galaxies twice as far away are moving away from us twice as fast,

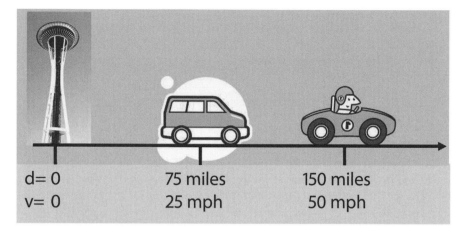

d= 0   75 miles   150 miles
v= 0   25 mph   50 mph

Figure 33.5. If all the vehicles leaving Seattle obey Hubble's Law their velocities would be proportional to their distances. Thus three hours before this image, all the vehicles were at the Space Needle. Since all distant galaxies are observed to obey Hubble's Law, they must all have been in one place at one time long ago. That time is the **Big Bang**.

and galaxies ⅓ as far away are moving away from us ⅓ as fast, as shown in figure 33.4. Hubble found a simple relationship between recessional velocity *v* and distance *d*: *v=Hd*. This is called Hubble's Law and *H* is called the Hubble expansion rate.

Let's see what this means with a down-to-earth example shown in figure 33.5. From atop the Seattle Space Needle, we observe a large number of vehicles leaving the city for a long weekend. Let's say that their velocities are found to obey the rule *v=Hd*. In this example, *H*=⅓. Thus the minivan is 75 miles away traveling 25 mph, and the sports car is 150 miles away (twice as far) traveling 50 mph (twice as fast). Assuming the vehicles' velocities are constant, where was the minivan 3 hours ago? It would have been at the base of the Space Needle since 25 mph times 3 hours is 75 miles. Where was the sports car 3 hours ago? Also at the Space Needle since 50 mph times 3 hours is 150 miles. In fact, all the vehicles on the road must have been in the parking lot of the Space Needle 3 hours ago because they all obey Hubble's Law, *v=Hd*.

What does this say about the universe? If all the galaxies obey Hubble's Law, as we observe they do, then in the distant past, every galaxy was in the same place at one time. We call that time the **Big Bang**.

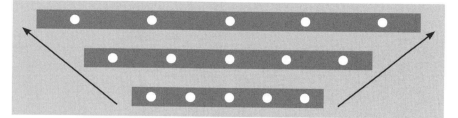

Figure 33.6. At the bottom is a strip of tape with coins (white dots) attached. As the tape stretches, the coins move apart, as shown in the middle and upper images. Coins twice as far apart move away twice as fast because there is twice as much tape stretching between them.

WHAT IS EXPANDING?

We should not interpret Hubble's Law as meaning that each galaxy is flying through space with a velocity that just happens to obey a certain equation. All these motions cannot be independent—there must be a mechanism that establishes the velocity of each galaxy. What is choreographing this cosmic dance? It is space itself. What is really happening is that space itself is expanding and carrying with it everything it contains.

Consider figure 33.6. Starting at the bottom, imagine a strip of elastic tape with coins attached to it. Stretch the tape farther and farther, as the figure shows. As the tape stretches, the coins move apart. Now, think of the coins as galaxies. Astronomers in each galaxy see all the other galaxies moving away. They see galaxies that are twice as far away move twice as fast, because there is twice as much tape stretching between them. They discover Hubble's Law. It isn't really the galaxies (coins) that are moving, it's the space (tape) between them that is expanding.

The stretching tape is expanding only in one dimension: length. Our universe has three dimensions of space and it's expanding in all three dimensions at the same time and at the same rate. We know this from the Cosmic Microwave Background (CMB) radiation that we will discuss in the next chapter and also from Type Ia supernovae. Figure 33.4 (and 33.7) include supernovae from all directions in the sky; since all these supernovae fall on the same line (same curve), the expansion rate must be the same in all directions.

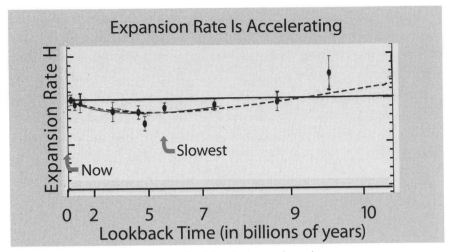

Figure 33.7. Horizontal axis is lookback time—how long ago a supernova occurred—which is related to its distance from Earth by the speed of light. Vertical axis is the expansion rate H, the supernova's velocity v divided by its distance d. H has changed over time. The time axis isn't linear. Plot is adapted from Riess et. al., www.arXiv.astro-ph/0402512v2, 31March2004.

Because observers in every galaxy see all other galaxies moving away from them, every observer has equal claim to being at the center of the universe. But in truth, there is no unique center, the universe appears the same at every point, in accordance with the Cosmological Principle.

The data in figure 33.4 are what was available before the Hubble Space Telescope started searching for Type Ia supernovae. Figure 33.7 includes all the prior data, now squeezed into the three points on the left, plus almost 200 additional Type Ia supernovae found by the Hubble telescope. The Hubble telescope found supernovae up to 7 times farther away and opened our eyes to a dramatic new understanding of the universe.

In figure 33.7, we plot *lookback time* horizontally and *H* vertically. Lookback time is how long ago a supernova occurred, which is related to its distance from Earth—light from a supernova with a lookback time of 1 million years has traveled 1 million light-years to reach us. The ratio v/d equals *H*, the Hubble expansion rate. H_0 is the expansion rate today (at $z=0$) as indicated by the horizontal line in the figure. We see from figure 33.7 that the expansion rate *H* has been different in the past.

Plotting the data this way allows us to examine the expansion rate more precisely. Points above the horizontal line correspond to faster expansion rates, while points below that line indicate slower expansion rates. Today is at the left edge of the chart. The nearest and most recent supernovae are toward the left. The most distant and oldest supernovae are toward the right. The farther we move to the right, the farther back in time we are looking, hence *lookback time*. The data show the expansion rate has changed over the last 10 billion years. The slowest expansion rate occurred roughly 6 billion years ago. Since then, the expansion rate has increased.

EXPANSION IS ACCELERATING

The supernovae data show that the expansion rate is accelerating. When this was first published in 1997, it shocked most scientists. Almost everyone expected exactly the opposite. Why? Well, all galaxies pull on one another through gravity. If galaxies were moving slowly, gravity would eventually bring them to a stop and then force them to start falling back toward one another. This is like a ball tossed upward; gravity slows its ascent and eventually brings it back down. If that happened, one day all the galaxies would all crash into one another in a *Big Crunch*. Even if galaxies were moving too fast for gravity to bring them to a stop—if they were moving faster than the escape velocity—gravity should still slow them down somewhat.

In fact, it does seem the expansion rate is slowing down in the right half of the figure. Moving from the oldest supernovae at the far right toward the more recent ones in the middle, the expansion rate appears to decrease. The data there are not definitive; at these vast distances, fewer supernovae are found and the measurements are less precise, as indicated by vertical lines through each data point. Nonetheless, there does seem to be a downward trend from right to middle, reaching a minimum about 6 billion years ago. The big surprise is what has happened since then. From middle to left, from 6 billion years ago to today, the expansion rate has been increasing. Something must be pushing the universe apart.

Perhaps this acceleration is due to some form of negative gravity. We attribute this surprising result to a new phenomenon dubbed *dark energy*, which is discussed in chapter 36.

From everything that astronomers have observed so far, the universe overall will continue expanding forever, and at an ever-increasing rate.

WHAT IS NOT EXPANDING?

But not everything is expanding. The Milky Way, the solar system, and Earth are not expanding. We humans are not expanding (at least not for cosmic reasons). All of us are held together by forces stronger than the expansion of space. At the current expansion rate, the diameter of Earth's orbit would increase by 70 feet per year. The Sun's gravity will continue to overcome this expansion, as it has for nearly 5 billion years. This expansion is much weaker than the gravitational forces that hold together Earth, our solar system, the Milky Way, and our own galaxy cluster, the Local Group. The expansion is also much weaker than the electric forces that hold together humans, molecules, and atoms.

Objects at the edge of our observable universe are moving away from us at the speed of light. To be more precise, each year the space between us expands by one light-year. Einstein said nothing can move through space faster than the speed of light, but space itself can expand at any speed without conflicting with Relativity. As the expansion of the universe accelerates, anything not already in our observable universe, like the monster in figure 29.3, cannot move fast enough to overcome the expansion of space, and thus will never get closer to us.

In the future, distant galaxies will move away so rapidly that they will move out of our observable universe. The radius of our observable universe—everything we can see—grows by one light-year per year. But very distant galaxies will move away even faster; they will exit our observable universe and will disappear from view. There will be less and less matter in our observable universe even though its physical size will increase. But nearby galaxies, and everything within them, will continue to remain at their current distances, or less.

EINSTEIN'S EXPENSIVE DIVORCE

When Einstein married Maleva Maric in 1903, they were poor outcasts deeply in love. She had failed to graduate from the Polytechnic and he had failed to find academic employment. But in 1909, Einstein got his first academic job and rapidly rose through the professorial ranks. Within five years he held one of the world's most prestigious professorships, and was becoming a celebrity. Bohemian intellectuals were in fashion and it is said ladies considered him dashing and wickedly exciting. Suffice it to say, Einstein did not win a Nobel Prize in Fidelity.

Maleva resented all this. She had sacrificed her own opportunity for academic success to support his career. Now he was cavorting while she was home caring for his children. Serious friction developed between them that only served to drive Albert farther from her—he strongly resented any attempt to limit his personal freedom. Their relationship became barely civil.

Starting in 1912, Einstein began an affair with his first cousin Elsa. She was happy to take Albert as he was—philandering, obsessed with physics, but famous. She knew his work would always be his first love.

The two women were completely different. Maleva had been Albert's intellectual colleague. Elsa liked to cook, and once said she didn't really know what Albert did, but she did know he was a famous "physician."

In 1914, Albert delivered a written ultimatum to Maleva demanding that she agree to do his laundry, serve him meals in his room, keep his room tidy, leave his desk untouched, not expect any intimacy, and not talk to him if he didn't wish it. After some consideration, Maleva declined.

Einstein begged Maleva for a divorce, but she adamantly refused. Finally, in 1918, he made a remarkable offer. He would give her half of his academic salary plus she would get all his Prize money if he received a Nobel. After painful deliberation, she finally acceded. Maleva and Albert were divorced, and he promptly married Elsa.

In 1922, Albert kept his word and turned over to Maleva the entirety of his Nobel Prize receipts—over $500,000 at current valuations.

34

CMB: the First Light

The *Cosmic Microwave Background* (CMB) radiation fills our universe. There are a billion times more CMB photons in the universe than atoms. The CMB is the first light of our universe and it has a grand story to tell about what our universe was like when it was very, very young.

In a 1948 paper, Ralph Alpher, Hans Bethe, and George Gamow said that if the universe had begun with a *Big Bang*, evidence of the primordial fireball would still be detectable today. Like radiation from any hot object, the light from this fireball would have a specific energy distribution, a *Planck black-body spectrum*. The fireball's light (what we now call the CMB) would have been redshifted as the universe expanded, and they estimated its equivalent temperature would now be 25 degrees above absolute zero, 25 K (–415 °F). The idea was very well-conceived, but the temperature they computed was too high because nuclear physics was not well understood at that time.

A bit of physics folklore goes along with this important paper. The real work was done by Alpher and Gamow, but when the paper was ready to publish, Gamow decided to add Bethe as a coauthor so the author list would sound like *alpha, beta, gamma—a, b, c* in Greek. Additional calculations were performed at the U.S. National Bureau of Standards by Alpher and Robert Herman on a newfangled device called an "electronic computer." In those days, a "computer" was usually someone who was adept with an adding machine. Gamow said Herman might have been added as a fourth author if only he had agreed to change his name to Delter. Such is physics humor.

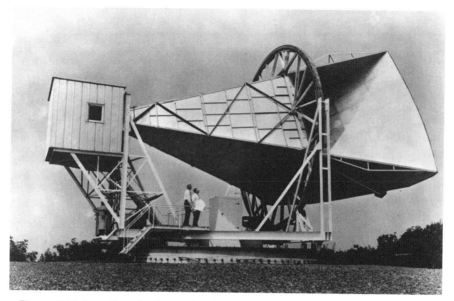

Figure 34.1. Arno Penzias (1933–), Robert Wilson (1936–), and the microwave antenna that first discovered the Cosmic Microwave Background radiation (CMB), strong evidence for the Big Bang theory. Image courtesy of NASA.

NOBEL OR NO BELL

In 1964, American physicists Arno Penzias and Robert Wilson of Bell Labs were working to improve microwave telecommunications. But they ran into a constant, static background interfering with their transmissions— there was noise on Ma Bell's phone lines. They methodically searched for the source of this annoying static, even cleaning up pigeon droppings from the inside of their large antenna, shown in figure 34.1. Eventually, with the help of Robert Dicke of Princeton, they became convinced that the static came from the universe itself—Penzias and Wilson had discovered the CMB, the afterglow of the *Big Bang*. The CMB is now microwave radiation at an equivalent temperature of 2.725 K (–455 °F).

Dicke was an expert cosmologist who was actively searching for the CMB, but did not have the right equipment. Penzias and Wilson were not looking for it and didn't even know what it was, but they had the right antenna. For their discovery, Penzias and Wilson shared the 1978 Nobel Prize in Physics (but not Dicke—*c'est la vie*).

CMB INFORMS COSMOLOGY

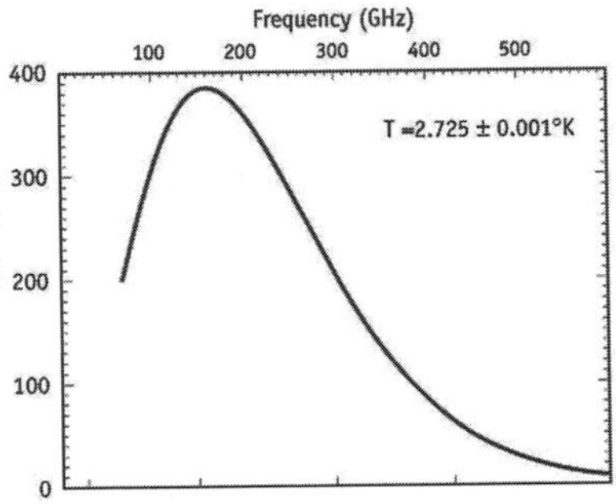

Figure 34.2. The CMB energy spectrum matches a Planck black-body spectrum with a peak temperature of 2.725 K.

The energy distribution of the CMB, shown in figure 34.2, is the most perfect thermal spectrum ever measured. Recall from chapter 16 that Planck solved the "ultraviolet catastrophe" by suggesting that radiation from hot objects is quantized. Planck showed that the intensity of radiation at each frequency has a very well-defined shape, called the *Planck black-body spectrum*. The CMB conforms to Planck's prediction to within the observational precision of 1/2000TH of 1%.

Another remarkable characteristic of the CMB is its uniformity across the entire sky. The difference in CMB temperature between any two regions of the sky is no more than one part in 100,000 (after adjusting for Earth's motion).

How is this possible? Normally, objects come to the same temperature by sharing heat through contact or radiation. Hotter objects transfer heat to colder ones until they equalize. But how can this happen in a vast universe? How can a region 13 billion light-years north of Earth have so nearly the same temperature as a region 13 billion light-years south of Earth? It would take at least 26 billion years for those two regions to send energy to one another, but the universe is not that old. The only explanation for this temperature uniformity is that our entire universe was once completely contained within a very small volume, making it easy for all of its parts to share their heat evenly.

Before the discovery of the CMB, many physicists had opposed the *Big Bang* theory. When the CMB was discovered, they struggled to find alternative explanations for its origin. But they all failed. Only the *Big Bang* theory is able to explain the CMB.

> The CMB provides strong evidence for the Big Bang theory.

The Big Bang theory states that when our universe began it was very small and very hot. Initially, protons and electrons were too hot to form atoms, and the universe was filled with unpaired, charged particles, a state of matter called *plasma*. Plasma is what we see in a flame. You can't see through a flame because plasma absorbs light: it's opaque. In the beginning, light was trapped within the opaque plasma.

As the universe expanded, it cooled down. Eventually, protons and electrons became cool enough to combine to form neutral hydrogen atoms. This ended the plasma era, making the universe transparent for the first time and releasing the CMB. These photons have been flying across the cosmos ever since. The CMB is the first light of our universe.

The CMB was released when our universe was 380,000 years old and everything in it had a temperature of 3000 K. Since then, the wavelength of the CMB has been stretched by a factor of 1092, dropping its temperature to 2.725 K. Therefore, since the CMB was released, the universe must have expanded in each direction of space by the same factor of 1092.

Many ground-breaking experiments have meticulously measured the CMB, including balloon experiments flown around the South Pole. The latest and greatest experiment is NASA's Wilkinson Microwave Anisotropy Probe (WMAP), a satellite launched in 2001.

These experiments have found minutely small, but very important, variations in the CMB. Color Plate 30 shows the results of WMAP's observations. This is an image of the universe when it was 1/36,000TH of its current age—it is a baby picture of our universe.

Comparing today's universe to a 100-year-old person, Color Plate 30 is its picture 24 hours after conception. We can already see "arms and legs." The red areas are where the energy density was just slightly higher; these grew to become galaxy clusters. The blue areas are where the density was slightly less; these are now great voids. What does the CMB tell us? An amazing amount, actually. The data are analyzed by comparing all pairs of directions in the sky, creating a *power spectrum* shown in figure 34.3. Basically, this is a precise measurement of how "clumpy" the image is. A power spectrum measures how many clumps there are of each clump size. The power spectrum of an oil painting would tell us the size of each brush the artist used and how often it was used. The location and height of the peaks reveals much about our universe.

WE LIVE IN THE GOLDEN AGE OF COSMOLOGY

WMAP achieved one of the most important advances in the history of cosmology. A generation ago, scientists knew few quantitative facts about our universe as a whole. Some thought the universe was infinitely old while others thought it was 2 billion years old. Two to infinity is quite a broad range—they really didn't know. Now, thanks in great part to WMAP, we have measured the age of our universe to 1% precision.

What CMB Tells Us	
Age of universe	13.7 Billion Years
Global curvature	0, space is Euclidean
Normal matter	4.5% of total energy
Dark matter	22% of total energy
Dark energy	73% of total energy

NASA WMAP (7-year data) measured to 1% precision.

The age of the universe is determined primarily by combining the expansion rate with the amount of expansion that has occurred since the

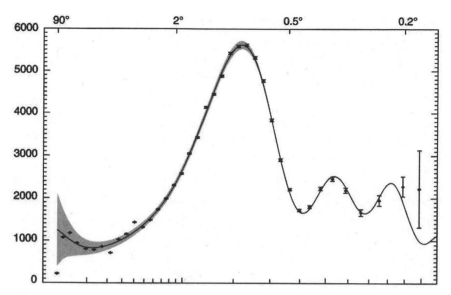

Figure 34.3. The CMB power spectrum. The curved line is the prediction of the Inflationary Big Bang theory. The dots represent measured data with the vertical lines indicating the instrumental uncertainties; many points are so precisely measured that the uncertainties are too small to see. The agreement between theory and experiment is outstanding. Plot from NASA WMAP analysis of 5 years of data collection.

CMB was released at *z=1091*. The CMB analysis tells us that, to a precision of 1%, 13.7 billion years have passed since CMB release, call that TAFTER. This means TAFTER is between 13.56 and 13.84 billion years. To this, we should add how old the universe was when the CMB was released, call that TBEFORE, which the analysis indicates is 380,000 years ±3200 years. Because TBEFORE is 360 times smaller than the uncertainty in the TAFTER (0.38 million vs. ±137 million), adding TBEFORE doesn't change the total within our current level of precision.

While space and time are curved near stars and galaxies, General Relativity allows for the possibility that the global curvature of the universe is positive, negative, or zero. Global curvature depends on the universe's average energy density, and for nearly a century, cosmologists struggled to find out what that was. Would the universe expand forever? Would it eventually re-collapse? Now we know.

The position of the main peak in the CMB power spectrum determines the universe's energy density. With the peak at about 1 degree, the measured density is consistent with zero curvature, within a precision of 1%. This means that the large-scale geometry of our universe is Euclidean.

The CMB data also determine the energy in photons and neutrinos. The amount of energy now in photons is very small, 0.005% of the total energy. If neutrinos had significant masses, their contribution might be important, as some scientists once proposed. But the CMB shows neutrinos now contribute less than 1% of the total energy. This means that most of the energy must be in electrons, protons, and neutrons. Right?

Wrong!

The ratio of even and odd peak heights in the power spectrum measures the normal matter density. It is startling that only 4.6% of all the energy in the universe is in atoms, according to the CMB. Less than $1/20^{TH}$ of the energy of the universe is in the form of normal matter—stuff like us—protons, neutrons, and electrons. Again we discover that we are made of rare and precious ingredients.

These data show that 95% of all the energy is in two newly discovered, exotic forms: *dark matter* and *dark energy*. Clearly, these are now very hot topics. The third peak in the power spectrum tells us how much dark matter exists. All the remaining energy, not in normal matter or dark matter, must be in the form of dark energy.

The next two chapters discuss the dark side.

SILK STOCKINGS SAVE DAD

During World War II, my dad, Oreste Piccioni, was assigned to the Italian patent office and was put in charge of reviewing patent applications for perpetual motion machines. It was a cushy job. Perpetual motion machines are the modern version of alchemists' claims of turning lead into gold. The "inventors" promise limitless, free energy from machines that run forever without consuming any fuel. Creating energy from nothing violates the fundamental principle that energy is conserved—its total amount never changes. Thus it was relatively easy to spot design errors and reject each application. (The U.S. Patent Office has made this process even easier by requiring that working models accompany all perpetual motion machine applications. So far they've received none.)

In 1943, Italian military leaders staged a coup and the new government surrendered to the Allies. The German army immediately took control of all of Italy, except the most southern portion that was occupied by the Allies. Conditions for most Italians went from bad to worse.

Dad joined a group of about a dozen young men who intended to escape from the Germans by hiking through the mountains to the south and crossing to the Allied side. Unfortunately, they were captured by a German patrol and two were immediately shot. The rest were taken to a prison camp where they were told they would all be executed or sent to labor camps in Poland. As dad said, "If German soldiers said they were going to shoot you, you could believe them."

The father of one prisoner turned out to be a man of influence and wealth. He was able to negotiate with the prison camp commandant. In addition to the usual money, cigarettes, and chocolates, he offered silk stockings, items much prized by certain women and unavailable during the war. With this contraband, he was able to buy the freedom of a handful of prisoners, including my dad. The rest were sent to Poland. Some managed to survive the labor camps, some didn't.

Dad returned to Rome and his famous muon experiment. He also started building radios for the underground, until the Allies liberated the city on June 5, 1944.

35

Dark Matter

Dark matter is a major component of our universe that is largely hidden from our view. We call it "dark" because it does not emit light and does not seem to interact with anything, except through gravity. Dark matter doesn't participate in the strong or electromagnetic forces as normal matter does. Whether or not dark matter participates in the weak force hasn't yet been established.

Why dark "matter"? From the CMB data discussed in the last chapter, we know this mysterious dark entity is a form of matter because its velocity relative to its surroundings is much less than the speed of light, therefore it cannot be any form of radiation. To emphasize that, it is often called *cold dark matter*. Why cold? Temperature is associated with the energy of atomic motion. Hot matter has high velocity and high kinetic energy. Cold matter has low velocity and low kinetic energy. The CMB tells us dark matter is cold.

The first hint of dark matter came 75 years ago, yet it remains a mystery to this day. We know more about what it isn't than what it is. While its composition and characteristics are unknown, we are quite sure that:

- Some form of cold, invisible matter is exerting huge gravitational forces on the normal matter and energy that we do see.
- This mysterious substance is not made of normal atoms nor their constituents: protons, neutrons, and electrons.

ARE WE SURE IT EXISTS?

The saga of dark matter began in 1933 when Caltech astrophysicist Fred Zwicky discovered that galaxies in clusters are moving much faster than expected. Newton's laws, which are quite adequate for this purpose, provide a powerful relationship for the orbit of a smaller body around a larger body of mass M. In natural units, $rv^2=M$, where v is the smaller body's velocity and r is its distance from the center of M. Zwicky measured v and r for many galaxies in several clusters, and from these he computed the required mass of each cluster. He found that the mass needed to account for the galaxies' high speeds was far greater than the total mass observed in the cluster—about 10 times greater. He concluded that most of the mass in a cluster must be invisible.

Scientists really didn't know what to do with this, and Zwicky's discovery languished for four decades. Finally, in the 1970s, American astronomer Vera Rubin discovered that galaxies are rotating too fast for the amount of normal matter they contain. Using the same logic as had Zwicky, Rubin measured the orbital speeds of stars at various distances from galaxy centers and found that much more mass must exist on the outskirts of galaxies than is visible.

More recently, studies of gravitational lensing, the bending of light by massive bodies discussed in chapter 22, found more evidence for invisible matter. The observed bending of light indicates that far more matter is present in galaxies and clusters than is visible.

Also, the CMB data discussed in the last chapter show the universe contains 6 times more matter than exists in protons, neutrons and electrons.

More evidence for the existence of dark matter comes from computer simulations that show the density variations in the early universe, evident in the CMB, are too small to have developed into galaxies without the additional gravity of dark matter. Figure 35.1 shows a computer simulation of matter condensing over the life of the universe, assuming the density variations and the amount of dark matter found in the CMB analysis. The end result, the right-most image, matches well with the large-scale structure we now observe in the real universe.

CAN IT BE NORMAL MATTER THAT WE CAN'T SEE?

Are we sure dark matter is not normal matter that we just cannot see? In Chapter 39, we'll discuss the evolution of our universe and we'll discover that the production of certain elements is very sensitive to the density of protons and neutrons. If dark matter were composed of protons and neutrons, the cosmic abundance of deuterium and other light elements would be rather different than what is actually observed. The observed abundances are consistent with the amount of normal matter we do see, but not with 6 times as much.

The CMB is also inconsistent with dark matter being some invisible form of normal matter.

OK, IT'S SOMETHING NEW. WHAT IS IT?

Physicists have been more successful coming up with clever names than clever answers. Some say dark matter is WIMPs: Weakly Interacting Massive Particles. Not to be outdone, others say dark matter is MACHOs: Massive Compact Halo Objects. Still others have proposed that for every existing fermion there is an as-yet-undiscovered boson, and vice versa. This prolific proposal is called *Supersymmetry*. Supersymmetry predicts hundreds of new particles, none of which have yet been found. (Politicians once promised a chicken in every pot. Supersymmetry promises a new particle for almost every experimental physicist. It's the Theory of Eternal Employment.) These new particles should decay to whichever particle of their type has the lowest mass, which is dubbed the LSP, Least-massive Supersymmetric Partner. Perhaps dark matter is made of LSPs that might be WIMPy. All these speculations are fine, they provide models for experimental searches, but our knowledge will not really be advanced until experiments actually capture some dark matter.

Color Plate 31 is a very interesting image of the Bullet Cluster. Here, we see the aftermath of the collision of two galaxy clusters that are 3.4 billion light-years away. The larger cluster on the left and a smaller cluster on the right passed through one another and are now moving apart. The

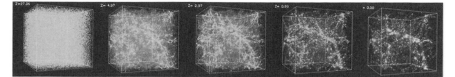

Figure 35.1. A computer simulation shows matter (white) condensing during the evolution of the universe. The five snapshots in time start at the Big Bang on the left when matter was very evenly distributed, and go through to today on the right, where we see a structure called the cosmic web. Images are scaled to compensate for the expansion of the universe. Images by A. Kravtsov and A. Klypin, National Center for Supercomputer Applications, Center for Cosmological Physics, U.S. National Science Foundation.

mass of the larger cluster is 1000 million, million *Msun*, and the clusters collided at a velocity of 10 million mph. This collision was perhaps the most violent event since the Big Bang itself.

The diffuse blue areas in this image represent calculated dark matter haloes. Because we cannot directly see dark matter, the amount and extent of each dark matter halo is computed from the observed amount of gravitational lensing (of light from more distant sources) caused by each cluster. The two red patches between the clusters are plasma clouds—vast oceans of charged particles (normal electrons and nuclei). Before the collision, each cluster was embedded in its own plasma cloud, as is normally the case.

Measurements show that the plasma clouds contain far more mass than all the galaxies in both clusters combined, and also that the dark matter contains far more mass than the plasma.

The image shows that the galaxies passed through one another with very little interaction because there is so much empty space between galaxies. Like two hails of bullets fired toward one another, most bullets sailed on by without hitting any on-coming bullets. While there may have been some near misses, there probably were no head-on collisions between galaxies. However, the plasma clouds did not sail on by. Charged particles interact vigorously and these interactions slowed down each plasma cloud, causing them to lag behind their clusters. The plasma cloud on the right displays a *shock front*, common in such violent events.

That leaves the dark matter. Unlike the galaxies, each dark matter halo is diffuse like the plasma, yet neither halo interacted with the plasma, the galaxies, or even the other dark matter halo. Dark matter seems quite inert, except for the gravitational attraction it exerts.

Color Plate 31 shows that dark matter cannot be made of normal atoms and particles.

DARK MATTER CAN'T CLUMP

Another important point: dark matter does not seem to clump as normal matter does. When a gas cloud collapses due to its self-gravity, the gas gets hot. The atoms move faster and exert pressure that can stop the collapse. Normal matter is able to radiate energy through the electromagnetic force. This allows normal matter to reduce its temperature and pressure, and thereby continue collapsing to form stars and planets. It seems dark matter cannot cool itself by releasing energy and is not able to collapse to nearly the same density as normal matter. Apparently, this is why dark matter is detected diffusely surrounding galaxies and galaxy clusters.

We believe that large-scale structures—galaxies and galaxy clusters—began with the accumulation of dark matter. Because there is so much more mass in dark matter, the attraction of its gravity was the most important factor in bringing material together to form large structures. Once dark matter accumulated, normal matter fell toward the center of dark matter's strong gravitational field. Ultimately, with its ability to cool by radiation, normal matter "out-clumped" dark matter and collapsed to a much denser state. Dark matter created fertile sites where normal matter condensed to form clusters, galaxies, stars, and planets.

Numerous experiments are currently underway attempting to catch and identify pieces of dark matter. We wait hopefully, but for now, dark matter remains a dark mystery.

FEYNMAN AT LOS ALAMOS

Feynman was renowned for his great sense of humor. Over dinner one evening, he told me this story.

During World War II, Feynman worked on the Manhattan Project, which was conducted at the newly-established, top-secret lab at Los Alamos, New Mexico. Security at the lab was supposed to be the U.S. Army's very best, but there were a few holes that Feynman delighted in highlighting. One day, Feynman took a walk around the compound hoping fresh air and solitude would inspire a new approach to a technical problem. As he walked along the inside of the security fence, he noticed that a hole had been cut in it. He went to the guard gate and reported this breach to the military police, but they didn't believe him and replied, "That can't happen here." To prove his point, Feynman walked back to the hole in the fence and went through, exiting the secure compound. He then walked back to the guard gate and said "See. I got out through the hole I told you about. Check your log. You didn't sign me out through the gate. The only other way out is through the fence." The guards didn't get it. So Feynman logged in, went back to the hole, exited, and returned to the guard gate—two more times. Eventually, they believed him. They dispatched soldiers to repair the fence and put Feynman's name on a list of trouble-making physicists.

FEYNMAN AT CALTECH

Whenever Feynman gave a lecture at Caltech, he filled the room, regardless of his topic. About 200 attended his class on Quantum Electrodynamics, although only 20 of us were registered students—the rest were other members of the Caltech physics department and professors from nearby universities. All came to experience the magic of an outstanding mind.

36

Dark Energy

Almost three-quarters of all the energy in the universe is in a mysterious form we call dark energy. Whatever it is, we know its negative gravity is dominating the universe and determining its fate. By negative gravity, we mean gravity that pushes things apart, in contrast with normal gravity that always pulls things together. Because of dark energy, the universe will expand forever, and expand at an ever-increasing rate. Some call this the *Big Rip*.

The primary evidence for the existence of dark energy comes from two sources: (1) the cosmic microwave background (**CMB**) radiation, which we discussed in chapter 34; and (2) the accelerating expansion of the universe that is observed using Type Ia supernovae, which we discussed in chapter 33.

The Type Ia supernovae data shown in figure 33.7 demonstrate that dark energy began to dominate all other forms of energy about 6 billion years ago. Before then, the attractive gravity of matter, both normal matter and dark matter, dominated and was gradually slowing down the expansion of the universe. In the last 6 billion years, the negative gravity of dark energy gained the upper hand and is now accelerating the expansion.

Physicists have some confidence they know what dark energy is, though we cannot prove it yet.

DARK ENERGY AND VIRTUAL PARTICLES

Recall the virtual particles of Quantum Mechanics discussed in chapter 17. Always and everywhere, particle-antiparticle pairs spontaneously flash into and out of existence. This is a real effect that has been extensively and precisely confirmed. One consequence of virtual particles is that any volume of space, even "empty" space, contains energy. Even though virtual particles lead the briefest of lives, every cubic foot of space will always contain some number of virtual particles. Thus it's not unreasonable that the average amount of energy per cubic foot is greater than zero, even in empty space. Not unreasonable, but we don't yet have an effective physical theory that allows us to calculate the amount of this energy. Our current theories say that the energy of empty space due to virtual particles is absurdly huge—these theories are clearly wrong. A deeper understanding of this issue awaits the completion of a theory called Quantum Gravity that we hope will successfully combine Quantum Mechanics and General Relativity.

Meanwhile, it seems inescapable that empty space must have energy. The Casimir effect [1] is a non-cosmological phenomenon that confirms this and that can be demonstrated in our laboratories.

If a certain volume of empty space does have energy E, then it turns out it must have a pressure P; in fact, to ensure energy conservation $P=-E$. Thus if the energy of empty space is greater than zero, its pressure must be negative. General Relativity says that a substance with energy and pressure exerts a gravitational force proportional to $E+3P$. Hence, the gravitational force of empty space should be negative, since:

$$E+3P = E-3E = -2E < 0.$$

THE BALANCE SHIFTED

When the universe was younger and smaller, it contained less space and the negative gravity of space was less important. Also, when the universe was smaller, the density of matter (both normal and dark) was higher, and matter's positive gravity was more important.

Matter's positive gravity exceeded dark energy's negative gravity, and although the universe continued expanding, it's expansion rate was gradually decreasing. But as space expanded, the balance shifted. The ever increasing volume of space led to an ever increasing amount of dark energy and also to an ever decreasing density of matter. A time came, about 6 billion years ago, when dark energy began to dominate.

As space continues to expand, the density of matter will drop further and the amount of space, and the amount of dark energy, will increase even more. The global expansion of the universe will become exponential, increasing ever more rapidly, without limit.

But as we discussed earlier, galaxy clusters, and the Local Group in particular, will remain together due to the strength of their mutual normal gravity. As long as the galaxies in the Local Group don't separate, the amount of space between them will not increase, and neither will the amount of dark energy in this region.

IS DARK ENERGY THE COSMOLOGICAL CONSTANT?

Is dark energy the reincarnation of Einstein's "greatest blunder", his cosmological constant that we discussed in chapter 33? Many physicists would say yes and applaud Einstein for predicting dark energy 80 years prior to its discovery.

That is not my view. I see two important differences between dark energy and Einstein's cosmological constant. The first key difference is the physical effect. Einstein introduced his cosmological constant to counterbalance the gravitational attraction of matter and provide a static solution to his Field Equations. Dark energy provides neither balance nor a static solution; it is relentlessly driving the universe ever farther away from balance and stability. The effect of dark energy is nothing like what Einstein sought to achieve.

Secondly, dark energy and the cosmological constant differ in methodology and philosophy. When Einstein modified his Field Equations in 1917, he added the cosmological constant to the geometry side of his equations, making them:

$$G + \Lambda = 8\pi T$$

This means the effect is somehow due to the geometry of our universe. Energy, in all its forms, is properly on the right side of the equations, included in T:

$$G = 8\pi(T_{mat} + T_{rad} + T_{darkm} + T_{darke})$$

Here, the terms on the right represent the contributions to the total T from normal matter, radiation, dark matter, and dark energy. Of course, it ultimately makes no computational difference whether one adds something on the left or one subtracts that same something from the right. But inserting an unknown geometry effect is very different philosophically from including the energy of virtual particles due to the quantum nature of the micro-world. The first is an unexplained mathematical fudge factor, whereas the second is a working hypothesis with a solid physical motivation.

I see no need to find false reasons to praise Einstein. His many unquestionable, outstanding achievements certainly place Albert Einstein with Sir Isaac Newton as the two greatest scientific minds of all time. People can debate which one was greater, but that these two stand head and shoulders above all others is beyond debate.

That each was sometimes wrong merely adds a touch of humanity.

NOTES

[1] In 1948, Dutch physicists Hendrik Casimir and Dirk Polder observed a force between two closely-spaced metal plates. Metals do not permit the oscillating electric fields that photons require. Hence, virtual photons are excluded from the gap between the plates if their wavelength exceeds the gap. Thus the virtual particle energy in the gap is less than normal. In particular, there's less energy in the gap than on the opposite side of each plate. If, as we suspect, virtual particle energy is dark energy and has repulsive gravity, there will be more dark energy pushing on the plates from the outside than pushing on them from the gap side. Hence, the plates will be pushed together, exactly as Casimir and Polder observed.

37

Our Special Place in the Cosmos

The cosmos contains vast numbers of stars and presumably vast numbers of planets orbiting those stars. But our home is not just another rock in an immense universe. Earth really is a very special place. Let's discuss some of the things that make Earth special, starting with the wide-angle view and gradually zooming in.

Firstly, the Milky Way is a good galaxy to call home. Some galaxies emit indescribably enormous amounts of radiation from their centers due to voracious super-massive black holes. The radiation from some galaxy cores would not just sterilize a planet, it could destroy entire solar systems. Take another look at Color Plates 17 and 18. Our galaxy is far more sedate and less dangerous. Certainly, it is now. And, perhaps, it has always been relatively tranquil. Sagittarius A*, the central black hole in our galaxy, has a mass of "only" 4 million *Msun*, 3000 times less than the most massive black holes. This indicates that Sag A* has consumed much less matter and emitted much less radiation than many of its heavier cousins.

Our galaxy resides in a relatively placid group of over 40 galaxies. Within the Local Group, the Milky Way and Andromeda are the only major galaxies (see Color Plate 26 and figure 28.1); most of the others are much smaller. Thus it's likely that the Milky Way has never suffered a collision with another major galaxy, at least so far.

Our solar system is favorably positioned within our galaxy. We are far enough from the galactic center, 27,000 light-years, to avoid its hazards. As sedate as the Milky Way is compared with some galaxies, all galactic centers are dangerous places. The most massive black holes are in the centers of galaxies, as is the most intense radiation. Also, the density of stars at the center of a galaxy is 100 million times larger than in our neighborhood. Stars near the center of our galaxy are moving at up to 3 million mph. The combination of so many stars moving so rapidly makes galactic cores shooting galleries. It is much safer to be far from the core—but not too far. The outer reaches of galaxies have very little carbon, oxygen, and other elements life requires. These elements are made in very massive stars and there are far fewer of those out in the boondocks. Our solar system is in a very favorable neighborhood in the suburbs, not too close and not too far. Astrophysicists call this the galaxy's *Goldilocks Zone*.

Our solar system is also special in several other ways.

Firstly, our solar system began at a favorable time, after the most violent cosmic fireworks had subsided. In the first 9 billion years of our universe's existence, it experienced the explosions of the most massive stars, the greatest number of galactic collisions, and the heyday of quasars. By the time our solar system began 4.57 billion years ago, the mayhem had waned, the orbits of most of our galactic neighbors had stabilized, and our universe had greatly expanded, thereby moving many potential hazards far away from us. The Wild West phase of the cosmos was largely over.

Additionally, during the 9 billion years before our solar system, generations of stars created and dispersed the heavier elements (all elements other than hydrogen and helium). When our solar system formed, the heavy elements were sufficiently abundant to support life.

The next point may seem odd, but our solar system is also special because it contains only one star. Most stars have partners; they are generally in couples or threesomes. While stable orbits are the rule around a lone star, it is less likely that a planet will find a stable orbit in a multiple-star system. Without an extremely stable orbit, a planet cannot sustain life.

The Sun also has a favorable mass. If the Sun's mass had been ⅓ higher, it would have burned out before humans evolved. If the Sun's mass had been ⅓ lower, the chance of any planet being in the *habitable zone* of our solar system would have been 4 times smaller.

It's good to have a big brother. Earth is the largest of the rocky terrestrial planets, but it is only a pebble compared with Jupiter. Jupiter has over 300 times Earth's mass and 1000 times its volume. In fact, Jupiter's mass is more than twice the combined masses of all our solar system's other planets. Jupiter patrols the outer solar system, deflecting or consuming many of the asteroids and comets that might otherwise rain down on the inner solar system, thereby making our neighborhood a safer place.

It's good to have a little sister as well. Our Moon's mass is 26,000 times less than Jupiter's and about 80 times less than Earth's. Yet the Moon has been very important in Earth's development. The Moon is believed to have formed after a "rock" the size of Mars struck Earth. That rock was another *proto-planet* like the developing Earth, but 10 times less massive. The rock has been named Theia after the mother of moon goddess Selene, from Greek mythology. Theia's impact is estimated to have occurred 30 to 50 million years after the proto-earth formed (4.5 billion years ago). The collision was cataclysmic. On impact, Theia's kinetic energy was converted to heat and melted the entire Earth from surface to core, allowing iron and other heavy elements to flow down to Earth's center. Because the impact was at a grazing angle, a great deal of material was blasted off into space. Much of it settled into orbit around Earth and eventually accumulated to form the Moon.

As a percentage of its host planet, our Moon is by far the largest in the solar system. No other moon has nearly as much effect on its planet as our Moon has on Earth. It creates substantial tides that are believed to have played a role in the development of life in the sea and on land. The Moon also stabilizes Earth's rotational axis, thus providing consistent seasons and giving life a consistent habitat.

After Theia's impact, iron concentrated in Earth's core. The in-fall of the accreting materials that created Earth, as well as radioactive decays facilitated by the weak force, keep much of that iron molten to this day. A molten iron core gives Earth a much larger magnetic field than it would otherwise have, 100 times larger than those of the other terrestrial planets. Our magnetic field protects our atmosphere from the solar wind—the 1 million tons of charged particles the Sun spews forth each second. Earth's magnetic field deflects almost all these particles. By comparison, the very weak magnetic field of Mars provides no protection and any atmosphere Mars once had was eroded by the solar wind long ago.

Earth's location in our solar system may be its primary asset. We are in the middle of the habitable zone—close enough to the Sun for water to be above freezing and far enough for it to be below boiling. Taking Neptune as the last planet, now that Pluto has been demoted, the habitable zone occupies less than 1/100,000TH of our entire planetary system. We are right in the middle of the solar system's Goldilocks Zone.

Having a favorable average orbital radius is not enough. If Earth's orbit was highly elliptical, the oceans might freeze in "winter" and boil in "summer." How elliptical are typical planetary orbits? Excluding Earth and Pluto, the average ellipticity of planetary orbits in our solar system is 7% (with Pluto it would be 9%). This means the average planet's distance from the Sun varies from 7% less to 7% more than average throughout its "year." The amount of sunlight received by the average planet varies from 14% less to 14% more than average. If Earth's orbit were that elliptical, our surface temperature would vary from below freezing to near boiling. Fortunately, Earth's orbit is a more perfect circle, being 4 times less elliptical than our solar system's average.

Earth's gravity is also favorable for life. The gravity of Jupiter-sized planets captures hydrogen and helium, which are by far the most abundant elements in the universe, accounting for 99.8% of all atoms. A planet that is heavy enough to capture hydrogen and helium becomes a ball of gas without a solid or liquid surface. Earth is light enough and warm enough that hydrogen and helium gases float to the top of our atmosphere and drift off into space. However, hydrogen atoms that react to form water or other heavier molecules are captured by Earth's gravity. Conversely, a planet that is too light would not capture an atmosphere or retain liquid water. Earth's gravity is strong enough to capture gases heavier than helium and weak enough not to drown in the lighter elements. We are in another Goldilocks Zone in terms of planetary gravity.

TWELVE REASONS EARTH IS SPECIAL

1. The Milky Way is a relatively placid galaxy.
2. The Sun is in our galaxy's Goldilocks Zone.
3. Our solar system began at the right time.
4. Our solar system has only one star.
5. Our Sun has a favorable mass.
6. Jupiter protects Earth.
7. Our Moon stabilizes Earth's axis.
8. Earth's magnetic field deflects the solar wind.
9. Earth is in our solar system's Goldilocks Zone.
10. Earth's gravity is favorable.
11. Our atmosphere sustains and protects life.
12. Plate tectonics renews Earth's surface.

Earth's atmosphere provides an abundant supply of gases that sustains life. Our atmosphere also protects Earth's surface from multiple deadly hazards: the most dangerous ultraviolet solar radiation, high-energy cosmic rays, and the shower of micrometeorites that continuously rain down from space at lethal velocities of 25,000 mph and more.

Finally, Earth's plate tectonics plays a critical role in supporting life. The motion of tectonic plates—each being a large portion of Earth's crust—continually renews our planet's surface, building continents, maintaining the oceans' chemistry, balancing the distribution of carbon, enriching the soil, regulating Earth's temperature, and bringing metals up from the interior to the surface.

While there may be billions of billions of planets out there, and probably many that are habitable, ours is truly a rare and special home.

See Color Plate 32 for an image of this special place taken from lunar orbit by NASA astronauts on Apollo 8, in 1968.

SHOOTING POOL WITH FEYNMAN

Our family had a modest pool table when I was growing up. Feynman, a friend of my father, taught me the rudiments of the game.

Lesson 1: Grip the Cue Stick at its Center of Mass

The first pearl of wisdom was to make sure you hold the cue stick at its center of mass—the place where there is an equal weight in front of and behind your grip. This minimizes unintended torques that might deflect the stick during your stroke. The key is to find the center of mass of a tapered cue stick.

Feynman's solution is ingenious: (1) place your two index fingers horizontally, parallel, and at least one foot apart; (2) lay the cue stick over these two fingers; (3) slowly bring your fingers together while keeping them parallel and at the same elevation; (4) where your fingers meet will be the center of mass.

This will work with any smooth object—try it on a yardstick for fun. Feynman's scheme works because, at each instant in time, whichever finger is farther from the center supports less of the object's weight than the other finger. Hence, the friction will be lower on the farther finger and it will slide under the stick, while the greater friction on the nearer finger will keep it from sliding. Thus the finger that is farther from the center of mass will always move closer to it. Generally, one finger will move for a while and eventually stop, then the other finger will move, and so forth. When the fingers finally touch, you have found the center of mass.

Lesson 2: Putting Spin on the Cue Ball

"To understand this shot...you need to read chapter 18."
(He was referring to **The Feynman Lectures on Physics**, Volume I.)

38

Can We Save Earth?

In chapter 37, we discussed the wonderful habitat that Earth has provided for our evolution and survival. This habitat faces numerous threats from within and without, some near-term and some long-range. This chapter discusses several longer-range threats and what we might do to avert them.

ONLY MANKIND CAN SAVE EARTH

Despite our poor environmental record so far, humanity has a unique opportunity to save life on Earth, and even to save the planet itself.

Many criticize mankind for polluting the air, water, and soil; for ravenously consuming natural resources; and for causing the extinction of many species while driving countless others to the brink. While much of this criticism is well justified, I believe we humans will modify our behaviors in time to avert the most dire prognoses. My belief is not based on evidence of improvement (there isn't much), but rather on a general faith in humanity, and on an abundance of optimism. Our species may not be smart enough to avoid making mistakes, but we are generally smart enough to learn from the mistakes we do make. Of all the complex life on Earth, humans are the most adaptable and the most successful competitor in the Darwinian struggle for survival. We didn't rise to the top of the food chain by being inept.

Chapter 10 discusses how we can vastly improve energy generation and reduce the associated pollution. Here, let's address two long-range concerns of immense proportion: mass extinction from celestial bombardment and incineration of Earth's surface. Only mankind has the potential to save all life on Earth from these threats.

ASTEROIDS AND COMETS

Every hard surface in the solar system is covered with craters—scars from asteroid and comet impacts. In the early solar system, bombardment was so intense that life on Earth was impossible. Eventually, most of the solar system's material was swept up into the Sun and the planets, and most of the rest settled into stable orbits that avoid these massive bodies. While maturation has drastically reduced bombardment, it will never completely stop. As with almost everything else, small impactors far outnumber large ones. Fortunately, small asteroids and comets (called meteorites or shooting stars) burn up harmlessly in our atmosphere.

However, mid-sized impactors can cause massive, local damage. In 1908, a 150-foot-wide object exploded over Tunguska, Siberia. With a blast equivalent to 10 megatons of TNT, it destroyed 80 million trees over 800 square miles. Had it hit a major city, millions would have perished. Impactors of this size are estimated to hit Earth once every 1000 years, on average. In 1989, an asteroid 6 times larger passed through Earth's orbit and missed us by just 6 hours.

Large impactors can devastate the whole planet. Evolutionary biologists point to numerous mass extinction events in Earth's history—brief periods during which most plants and animals died and a substantial fraction of all existing species became extinct. Extreme volcanic activity and celestial bombardment are thought to be the main killers. Strong evidence exists that a 6-mile-wide asteroid hit the coast of Mexico's Yucatan peninsula 65 million years ago, causing the "K-T" extinction in which the dinosaurs and perhaps one-third of all existing species perished.

Impacts of that magnitude are estimated to occur once every 40 million years, on average.

Can mere mortals stop the sky from falling?

Yes! Well, mostly.

There are two types of impactors to consider: inner solar system bodies (primarily asteroids) and outer solar system bodies (primarily comets).

Large inner solar system bodies are relatively easy to find and track. There are now several active tracking programs, including NEAR (Near Earth Asteroid Research) at MIT that is funded by NASA and the U.S. Air Force. Once an impactor is found, its path is analyzed to determine if and when it will collide with Earth. Such collisions could be avoided by changing the impactor's speed or direction. Given enough time, perhaps decades, a rocket can intercept the impactor and gradually divert it from its collision course. There are several schemes for generating diverting forces, with the best choice depending on the impactor's composition.

Some asteroids are solid rock, while others are piles of loose rubble held together by their weak self-gravity. Smaller, solid asteroids could be deflected by conventional high-powered rockets, or by simply ramming them to knock them off-course. The larger asteroids require a sustained force over a very long time. For rubble piles, the force must be gentle enough not to separate the rubble into many small asteroids, all on virtually the same path—separated rubble could be as deadly as the original loose pile. Sustained forces could be generated by low-power nuclear reactors, by sails catching the solar wind, or by mirrors focusing sunlight. The key is having sufficient lead time for these forces to adequately change the asteroid's course. This should be possible for inner solar system bodies that remain in relatively stable, highly-visible orbits for millennia. We haven't yet perfected the required technologies, but the challenges seem manageable, given adequate and timely investment.

However, this is not a challenge we can successfully deal with at the last moment—we must avoid a "crash program."

Impactors from the outer solar system are more difficult. These bodies will most likely be comets—balls of ice and dirt. Typically, their orbits are highly elliptical, and they approach the inner solar system at very high velocities. With current technology, we may be able to detect them only a few months before impact—probably not soon enough. Much more advanced technologies must be developed in order to find these impactors sooner and rapidly generate large deflections.

INCINERATION BY THE SUN

If we successfully avoid extinction by bombardment, we will eventually face a temperature crisis.

The current concerns about global warming are very serious, but they pale in comparison to what will happen as our Sun ages. The Sun's luminosity is gradually increasing. In about 600 million years, Earth's surface temperature may be above the boiling point of water. Eventually, the temperature will rise above the melting point of lead (600 °F).

Can humanity save the planet?

Yes!

We could move Earth farther from the Sun. This could be achieved by steering a large asteroid into an orbit around the Sun that passes close to Jupiter and then close to Earth. Each time the asteroid passes in front of Earth, its gravity would pull Earth forward, modestly increasing our orbital velocity by transferring some of its orbital energy to Earth. The asteroid's energy would be restored each time it passes behind Jupiter and is pulled forward by the giant planet. Each orbital energy boost would move Earth a bit farther from the Sun. Over hundreds of million of years, we could keep Earth in a favorable orbit as the Sun brightens.

Again, we will only be able to avoid this disaster if mankind adopts a much longer planning horizon than we ever have before.

39

The Big Bang

By now, we have built the foundation needed to explore science's best theory of the evolution of our universe—the Inflationary Big Bang Theory. Figure 39.1 highlights this evolution, from the beginning of everything on the left to today on the right. We will call the beginning of time *t=0*, and will divide the entire history of the universe into three eras:

Evolution of the Universe	
The Beginning	t = 0
Era of Quantum Gravity	$t < 10^{-39}$ second
Era of Inflation	t in between
Era of Big Bang	$t > 10^{-30}$ sec. (*"1 tic"*)

These eras approximately correspond to the three zones of confidence in the 62-digit time scale of the universe discussed in chapter 28 and reproduced below.

40000000000000000.000000000000000000000000000 000000000000 0005 sec

The Era of the Big Bang is the gradual expansion and maturation of the universe. It began 13.7 billion years ago and continues through the

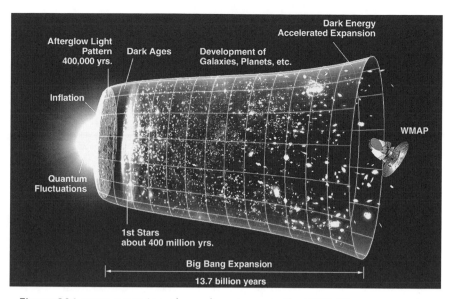

Figure 39.1. NASA overview shows key stages in the 13.7-billion-year evolution of the universe according to the Inflationary Big Bang theory. On the left is the creation of the universe from quantum fluctuations, followed by an era of very rapid growth, followed by more gradual growth, and ending on the right with the accelerating expansion.

present day. The Big Bang began 10^{-30} seconds after *t=0*. That is, it began a millionth of a millionth of a millionth of a millionth of a millionth of a second after the beginning. Rather than continue to deal with so many millionths, let's call that *1 tic*; it's much too brief to be a whole "tick-tock." While dark matter and dark energy remain mysterious, we believe they will eventually be incorporated into our existing physical theories as additional features, rather than requiring radically different theories. With that assumption, we have high confidence in our understanding of the basic physical laws that underlie the Era of the Big Bang. This confidence is based on our knowledge of elementary particles and on the two great pillars of 20TH century physics: Relativity and Quantum Mechanics. Over the last 100 years, both theories have been extensively tested with extraordinary precision over a very broad range of conditions.

Less certain is the Era of Inflation, beginning one billionth of a *tic* after *t=0* and ending at about *1 tic*. This is a period of hyper-rapid expansion.

There is a theoretical framework for understanding *Inflation*, but the details are still in flux.

Most uncertain is the Era of Quantum Gravity, the very beginning of time that spans only the first one billionth of *1 tic*. Here we know our current physical theories fail. We believe they fail because, in these extreme conditions, space and time are no longer smooth and continuous; thus our normal mathematical tools are not applicable.

One might well ask: "If we understand what happened after the first *1 tic*, all the way through the next 13.7 billion years, why bother with that infinitesimal time before? Why not just *fuggedaboudit*?" The answer is: It is exactly because we don't understand the first *1 tic* that makes it so exciting. What gets physicists out of bed in the morning is what they don't know. They are driven by the desire to discover nature's secrets. A thousand physicists are devoting their careers to understanding that mysterious, first, tiny *1 tic*.

We'll save speculative discussions about how the universe began and what came before, if anything, for chapter 40. Here, we begin just after the universe came into existence, as shown on the time line below.

ERA OF QUANTUM GRAVITY

↓

40000000000000000000.00000000000000000000000000000 000000000000 0005 s

What do we know about the beginning—the Era of Quantum Gravity? From the CMB analysis of chapter 34, we know with superb precision that the universe began 13.7 billion years ago. Our universe began as a fantastically small object, perhaps as small as the *Planck length*, less than a millionth of a millionth of a millionth of a millionth of the size of the smallest atom. Initially, it was also fantastically hot, probably over 100 million, million, million, million, million degrees. Our universe has been expanding and cooling ever since.

The name *Big Bang* is something of a misnomer. It was coined by British astrophysicist Sir Fred Hoyle who espoused a competing theory in which the universe existed forever, with no beginning and no end. Hoyle may have chosen to denigrate the opposition with a seemingly derisive name. (Perhaps you thought scientists were above this sort of thing?) "Big Bang" brings to mind an explosion of matter blasting through a pre-existing space. Actually, there was no pre-existing space or time. The most important aspect of the event at *t=0* is that our space and time came into existence. Matter did not explode out into space; space expanded and carried with it everything that existed.

We do not know much more about this era because our major physical theories fail here. Quantum Mechanics is a superb theory of the micro-world, of minute energies in minute distances. General Relativity is a superb theory of the macro-world, of immense energies and immense distances. But neither works when immense energies are concentrated within minute distances, such as at the beginning of time and at the center of every black hole.

Quantum Mechanics has a major flaw—it is not ***properly written*** in the language of tensors in four-dimensional, curved spacetime. Hence, its equations do not remain valid in all references frames, in any state of motion, and in any gravitational field. This deficiency is sometimes called a lack of ***general covariance*** or a lack of ***background independence***. For nearly a century, physicists have known that all physical laws must obey general covariance (must be ***properly written***). But this has never been successfully incorporated into Quantum Mechanics (or any of its derivative theories, such as String/M theory). Its equations work extremely well, but only when applied where gravity is weak and nearly uniform.

General Relativity also has a major flaw—it is not quantized. Its equations assume that energy, space, and time are continuous and smooth, like a ramp. For nearly a century, physicists have known that quantization is an essential feature of the micro-world, but this has never been successfully incorporated into General Relativity. Its equations work extremely well, but only when applied in the macro-world where the staircase steps are so small and so numerous that the staircases effectively become ramps.

Indeed, the problem is even deeper. Calculus, the mathematical foundation of all current physical theories, may not apply in this era.

Calculus assumes a continuous and smooth space and time in which its mathematical procedures can be sensibly implemented. Near the Planck length, our notions of continuous space and time seem likely to fail. Quantum Mechanics allows virtual masses to pop into and out of existence anywhere, anytime. General Relativity says that, at the Planck length, these virtual masses can achieve sufficient energy densities to implode into black holes that instantaneously evaporate. Spacetime becomes what John Archibald Wheeler called *quantum foam*, akin to the froth of an ocean wave crashing onto a rocky shore. At the scale of quantum foam, holes in spacetime are continually appearing, perhaps connecting distant points through *hyperspace*, and instantly vanishing. Even the number of dimensions of spacetime may not always be four. It seems at this scale, space and time may be quantized, as we found energy is quantized at the scale of Quantum Mechanics. Spacetime may exist only in discrete pieces that Italian physicist Carlo Rovelli calls "chunks of space." Perhaps nature allows 1 "chunk", or 2, or 170, but never 2½ "chunks."

Before we can understand more about this era, a new theory of physics must be developed to deal with these challenges. That theory has been named Quantum Gravity. It must incorporate the strengths of our current theories and overcome their weaknesses.

ERA OF INFLATION

↓

40000000000000000000.00000000000000000000000000000 000000000000 0005 s

By a billionth of *1 tic*, the universe is large enough and cool enough to emerge from the Planck scale. Spacetime is nearly continuous and calculus is now a useful tool. Now the Era of Inflation begins. During Inflation, the universe expands by an incredible amount, and at an exponential rate. We believe that during Inflation, the universe doubles in size every one millionth of *1 tic*. In less than *1 tic*, the universe grows by a factor of more than 10^{30}, perhaps vastly more. This expansion factor is at least 1000 times greater than the expansion factor during the next 13.7 billion

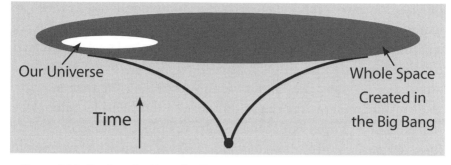

Figure 39.2. During the Era of Inflation, the universe grew spectacularly—more than 10^{30} fold. Here time ranges from t=0 at the bottom to t=1 tic at the top. In the next 13.7 billion years, the expansion has been "only" 10^{27} fold.

years. As figure 39.2 illustrates, the space created at *t=0* probably expands to become vastly larger than our observable universe.

Three very important things happen during the Era of Inflation.

Firstly, the geometry of our universe becomes Euclidean. Any initial curvature of space that may have existed before Inflation is reduced to virtually nothing. We can understand this by considering a small ball, like a ping-pong ball; it is clearly highly curved. Earth is also curved, but normally we don't notice its curvature because Earth is so large. The curvature of any ball decreases with the square of its radius—a ball 10 times larger has $1/100^{TH}$ as much curvature. The immense expansion during Inflation removes any initial curvature, "flattening" our universe, and making its geometry Euclidean.

This flattening is essential to creating a livable universe. If, immediately after Inflation, the overall curvature of space were $+10^{-50}$ or more, our universe would have rapidly and completely collapsed in a Big Crunch. If the overall curvature of space were -10^{-50} or less, our universe would have expanded so rapidly that no stars could have formed. Our existence depends on the overall curvature of space at the end of Inflation being virtually zero—differing from zero by no more than ±1 in the 50TH decimal digit! The spectacular expansion during Inflation creates this amazingly precise flatness.

Secondly, Inflation transforms an infinitesimally small ball, something immensely smaller than a single atom, into our entire universe.

It is easy for something small to have a uniform temperature (there's an "almost" coming). Its "pieces" are in such intimate contact that they cannot have different temperatures. Inflation explains why the CMB radiation has so nearly the same temperature everywhere: because at one time, it was all in essentially one place, close enough to share its heat energy.

And thirdly, here's the critical "almost." Quantum Mechanics says that everything, even zero, comes in shades of gray. The temperature throughout our pre-Inflation universe could not be exactly the same, but only "almost" exactly the same everywhere. The temperature and energy density are very slightly higher in some areas and very slightly lower in others, these variations are called *quantum fluctuations*. As our universe inflates from subatomic to about one foot wide, the quantum fluctuations also inflate and become the variations seen in the CMB in Color Plate 30. The regions with slightly higher density become galaxy clusters, and the regions with slightly lower density become great voids in our universe. Without this critical "almost", our entire universe would have the same density everywhere and there would be no stars, no planets, and no us.

During Inflation, space expands far faster than the speed of light. What? Didn't Einstein say nothing can travel faster than c, the speed of light? Actually, no *thing* can travel *through* space faster than c. But Relativity allows space itself to expand at any rate at all, even faster than light. As space expands, it carries with it everything it contains.

At the end of *1 tic*, the exponential growth of the Era of Inflation stops, and the energy which drove that growth is converted into all the particles of nature.

FRW EXPANSION EQUATION

Gravity now dominates quantum effects and the universe evolves according to Einstein's Theory of General Relativity and his Field Equations, $G=8\pi T$. From the CMB and other data, we know our universe is homogeneous and Euclidean; its density is nearly uniform and space is globally flat. For such a universe, Einstein's equations have a marvelous result—they have only one solution! Newton's equations have an infinite number of solutions for expanding universes and we'd never know which applies to ours. But because Einstein's equations unite spacetime with mass and energy, they have only one solution for a homogenous, Euclidean universe. This solution for the evolution of the universe was discovered by Alexander Friedman in 1922. It is used with a geometric description of homogenous spacetime developed by two mathematicians, American Howard Percy Robertson and Englishman Arthur Geoffrey Walker. Picking a letter from each name, this is called the FRW expansion equation, and considering that this is the equation for the universe, it is remarkably simple:

> For our universe,
> Einstein's Field Equations have only one solution:
>
> FRW Expansion Equation: $3H^2 = 8\pi e$

Here H is the expansion rate of the universe at any selected time and e is the energy density of the universe at that same time. With the FRW equation and measurements of H and e, we can compute the size of our universe at any time in the future, or at any time in the past back to *1 tic*. (The Schwarzschild solution of Einstein's equations is for a different problem: an isolated mass in a limitless, empty space.)

ERA OF BIG BANG

↓

40000000000000000000.0000000000000000000000000000000 000000000000 0005 s

The Era of the Big Bang begins at the end of Inflation, and continues to the present day. It is a period of more gradual expansion and maturation. As the universe expands, it cools and matter begins to condense into ever larger assemblies, as illustrated in figure 39.3. The structures in our universe start growing at the smallest scales and later proceed to larger and larger scales—the universe grows from the bottom up.

QUARKS COOL, CREATING NUCLEAR PARTICLES

↓

40000000000000000.000000000000000000000000000 000000000000 0005 s

We now move past a great many zeros on our time line, during which the universe keeps expanding and cooling. Eventually, at one millionth of a second after *t=0*, the temperature plummets to 10 million, million degrees (hot by human standards, but less than a billionth of a billionth of the temperature at the very beginning). This temperature is cool enough for quarks to stick together and form protons, neutrons, antiprotons, antineutrons, and hundreds of other particles and antiparticles that participate in the strong force. My father, Oreste Piccioni, helped discover some of the primary particles of antimatter.

ANTIMATTER VANISHES

↓

40000000000000000.000000000000000000000000000 000000000000 0005 s

At *t=1* second, the temperature is a mere 10 billion degrees. This is too cold for new particles to be created in collisions, as shown in figure 7.4. Without a continuing source, particles and antiparticles that annihilate, as shown in figure 7.3, are not replaced. Soon, all the antiparticles vanish. Why don't all the particles vanish as well? Almost all of them do, but fortunately, a very small fraction remains.

Generally, nature makes no distinction between matter and antimatter. This leads us to suspect that matter would disappear with the antimatter. If that happened, the universe would be devoid of normal matter and our story would never have begun. However, the lowly weak force saves the day. Only the weak force treats matter differently from antimatter. My Ph.D. thesis experiment explored this very small effect. During the first 1 second, the weak force allows a slight excess of matter to develop. As each antiparticle annihilates, it takes a particle of matter

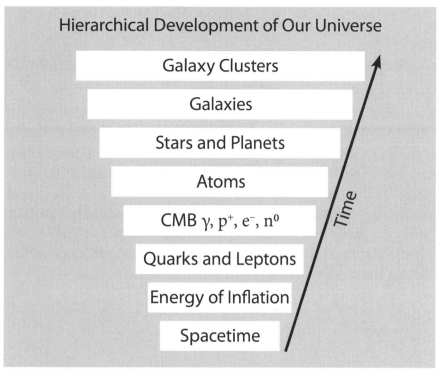

Figure 39.3. Our universe began at an infinitesimal scale. As time progressed, larger and larger structures formed.

with it. When all the antiparticles are gone, the slight excess of matter remains. That slight excess provides all the material from which galaxies, stars, and planets are made.

The amount by which matter exceeded antimatter was slight indeed. Every one billion antiprotons were outnumbered by one billion *and one* protons. After all the antimatter annihilated and only the matter remained, our universe contained 10^{89} CMB photons and 10^{80} protons, neutrons, and electrons. And that's all there is today.

How do we know whether it's the matter or the antimatter that the weak force chooses to favor? How do we know that we aren't made of antiprotons, antineutrons, and antielectrons? Maybe we are. However, we are the ones who chose the names of the particles, and we chose to call our stuff matter and call the other stuff antimatter. History is written by the survivors.

SYNTHESIS OF LIGHT NUCLEI

↓

40000000000000000000.00000000000000000000000000000 000000000000 0005 s

At *t=100* seconds, the temperature is 1 billion degrees. Protons (*p*) and neutrons (*n*) are now cool enough to stick together and form the nuclei of the light elements. These include deuterium (whose nucleus has one proton and one neutron: *1p1n*), normal helium (*2p2n*) and its other isotope helium–3 (*2p1n*), and just a dash of lithium (*3p4n*). There is not yet any carbon (*6p6n*) or oxygen (*8p8n*) in the entire universe. The Inflationary Big Bang theory predicts the abundance of the light nuclei and these predictions match the observed abundances very well, providing another strong confirmation of this theory.

FIRST ATOMS

↓

40000000000000000000.00000000000000000000000000000 000000000000 0005 s

At *t=380,000* years, the first atoms form. The temperature of the universe has dropped all the way down to 3000 K, about half the surface temperature of the Sun. Electrons and nuclei are now cold enough to combine and form neutral atoms. The universe is no longer filled with unpaired, charged particles that absorb light—the universe is transformed from the primordial plasma fireball to a transparent gas. No longer trapped by opaque plasma, the CMB radiation is released and has been flying through the cosmos ever since.

Because these CMB photons have had almost no interactions since the universe became transparent, the images astronomers take today of the CMB, such as Color Plate 30, show the universe as it was at *t=380,000* years, nearly 13.7 billion years ago.

FIRST STARS

↓
40000000000000000.0000000000000000000000000000 000000000000 0005 s

At *t=200* million years, the first stars form. Because the universe's density is still very high, the first stars are particularly massive and short-lived. Within their cores, they produce heavy elements, including carbon, oxygen, and iron. Recall that these elements are produced only in stars. Color Plate 33 is an image of the Rosette Nebula taken with filters that select light from certain elements. One filter selects oxygen atoms that are displayed in green. While oxygen comprises less than 1/1000TH of the atoms in our universe, you can clearly see there is a much higher concentration in the material ejected by dying stars.

SUN AND EARTH FORM

↓
40000000000000000.0000000000000000000000000000 000000000000 0005 s

At *t=9* billion years, our solar system forms. The Sun and the ~~nine~~ eight planets form at nearly the same time; see Color Plates 23 and 32. Our solar system condenses from a cloud of gas that was enriched with heavy elements by earlier generations of stars. The abundance of these elements in our Sun indicates it is probably a third-generation star.

From here on, we can more precisely determine events retrospectively; thus we will state how many years ago an event occurred, rather than how many years after the Big Bang. Once gaseous materials condense to form solids, their atoms are locked in fixed positions, allowing precise radioisotope dating of when the gas solidified. From such dating, we know that our solar system began solidifying 4.57 billion years ago, which is measured to a precision of 1% (±46 million years).

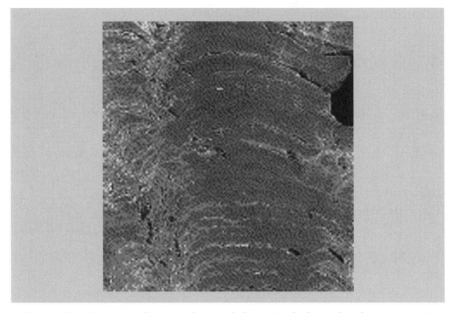

Figure 39.4. Stromatalites are layered deposits believed to have organic origin. The oldest stromatalite is estimated to be 3.46 billion years old.

THAT'S LIFE!

↓
40000000000000000.00000000000000000000000000000 000000000000 0005 s

At about *t=10* billion years (3.46 billion years ago), there is evidence of the emergence of life on Earth. Figure 39.4 shows a stromatalite believed to have been formed, layer upon layer, by single-celled, blue-green algae.

Amazingly, it took nearly 3 billion years for life on Earth to evolve from single-celled to multi-celled organisms. Multicellular animals first developed only about 600 million years ago. These were followed by the first fish 500 million years ago, and the first reptiles 300 million years ago. Mammals first appeared 200 million years ago.

The first humanoids evolved from primate ancestors about 6 million years ago. Our species, Homo sapiens, originated in Africa 200,000 years ago, and when the last Neanderthals died out 25,000 years ago, modern humans emerged as the last surviving subspecies of our genus.

Figure 39.5. Apollo 11 lands on the Moon in 1969. Image courtesy of NASA.

WE HAVE ARRIVED

↓
40000000000000000.00000000000000000000000000000 000000000000 0005 s

Finally, after a 13.7-billion-year journey from the beginning of our universe and the beginning of time, here we are today. See figure 39.5.

We live in the Goldilocks Zone of a solar system that is in the Goldilocks Zone of a grand galaxy in a universe with a perfect geometry. We are made of the rarest and finest ingredients that were billions of years in the making. Our atoms are the one in a billion that survived annihilation by antimatter, and one in ten of those that are in stellar systems, and one in 500 of those that are suitable for life, and one in a million of those that are on terrestrial planets.

40

What Came Before?

Up to this point, we have discussed well-established science.

Now let's consider how our universe may have begun and what may have come before. Without an established scientific theory capable of addressing the extreme conditions at very beginning of time, we now delve into speculation. Not wild, unreasoned speculation, but rather the educated guesses of several prominent researchers. There is no definitive proof for any of the ideas in this chapter and, to my knowledge, none of these ideas leads to a definitive falsifiable prediction, at least not yet.

HAWKING: NO "BEFORE" BEFORE

One proposal that cuts our speculations short is that there was no "before" before the beginning. Stephen Hawking proposes that time is an imaginary quantity, imaginary in the mathematical sense of being a multiple of the square root of –1. The three dimensions of space and one dimension of imaginary-time are then on a more equal footing, from a mathematical viewpoint. Space-imaginary-time is equivalent to the four-dimensional spherical surface of a five-dimensional ball. The beginning, $t=0$, is like the North Pole. Asking what came before $t=0$ is like asking what is north of the North Pole. Geometrically, Earth's surface at the North Pole is the same as it is anywhere else: round, smooth and continuous. However, at the North Pole, every direction within Earth's surface points due south. In the same sense, *t=0* is like any other point in Hawking's four-dimensional space-imaginary-time. Except that at *t=0*, every direction points forward in time; there are no points within the four-dimensional space-imaginary-time with *t<0*.

THE MULTIVERSE

All the following proposals assume a vast existence beyond our own universe: a *multiverse* that might be infinitely large and infinitely old, and might contain countless universes of which ours is only one. Our universe would exist in its own spacetime that is distinct and forever separate from the rest of the multiverse.

GUTH: THE ULTIMATE FREE LUNCH

A widely accepted view is proposed by American physicist Alan Guth, who conceived the theory of Inflation that has become an integral part of the Big Bang theory.

Guth says our universe came from nothing. Just as virtual particles can pop into existence by borrowing energy from the Bank of Heisenberg, Guth says that an extremely large, and very rare, quantum fluctuation could have created our universe. Guth envisions the spontaneous creation of a super-charged version of dark energy, which he calls an *inflaton,* that drives the exponential expansion of space during the Era of Inflation. After *1 tic,* the inflaton decays and its energy is converted into radiation and matter.

And the rest is, well, all of history.

Guth says, "The universe is the ultimate free lunch."

LINDE: ETERNAL INFLATION

Russian-American physicist Andre Linde independently conceived the theory of Inflation. He proposes the theory of Eternal Inflation in which quantum fluctuations, such as the one that created our universe, are continually occurring in various locations, at various times, as illustrated in figure 40.1. Recall from chapter 28 that the total energy of our universe is virtually zero. This means that whole new universes can be created from virtually nothing. Linde says new universes pop up here and there, now and then. Each might be somewhat different due to the details of how it starts. There might even be different laws of physics in each universe. Some might be habitable, but most probably would not. The different laws of physics are indicated in the figure by different shadings.

Figure 40.1. In Linde's theory of Eternal Inflation, new universes are continually forming and expanding through Inflation, much as ours did. Each universe could begin with different physical characteristics and even have different physical laws, as indicated by the various shades of gray.

LOOP QUANTUM GRAVITY:
COLLAPSE BEFORE EXPANSION

Loop Quantum Gravity (LQG) is the most promising effort, in my view, toward completing an effective theory of Quantum Gravity. Two of its leading proponents are Lee Smolin and Carlo Rovelli. Smolin is an American physicist at the Perimeter Institute in Ontario, Canada. Rovelli, an Italian physicist, is at the Center for Theoretical Physics in Marseille on the French Riviera (a tough gig, but someone's got to do it). LQG has made some progress in understanding the beginning. This theory proposes that spacetime is quantized, meaning there is a smallest "chunk" of spacetime that cannot be subdivided.

Calculations based on LQG principles, performed by Martin Bojowald of Pennsylvania State University, show that the universe was collapsing before the expansion that we call the *Big Bang*. In this scenario, called the *Big Bounce*, the universe was the smallest possible size, one "chunk", at $t=0$ and it was 4 "chunks" both immediately prior and immediately after. Bojowald's calculations also lead naturally to an era of very rapid expansion, similar to Inflation, following $t=0$.

One possible inference of the *Big Bounce* is that some cosmic body in another spacetime collapsed (perhaps a black hole), and our universe was born from that collapse.

Another possibility is that an entire universe collapsed to the smallest possible size, one "chunk", and then re-expanded, becoming our universe. What we call $t=0$ is the end of the collapse of the prior universe and the beginning of the expansion of ours. In this scenario, the universe might be cyclical, expanding, collapsing, and re-expanding over and over again. If this is true, the universe might be eternal. Life could not survive the collapse, but it would, in all probability, reoccur in every expansion phase. In that sense, life would be eternal as well.

SMOLIN: COSMOLOGICAL NATURAL SELECTION

I saved my favorite for last. This is the most interesting speculation I have heard and I hope it is true. It was conceived by Lee Smolin, and discussed in his fine book *The Life of the Cosmos*. Smolin calls this wonderful idea the theory of Cosmological Natural Selection. It starts where so much of physics does—with Albert Einstein.

> "What really interests me is whether God had any choice in the creation of the universe."
> — Albert Einstein, Nobel 1922

With what we now know, we can begin to answer Einstein's question. There seem to be about 20 choices—20 critically important parameters of a universe: the mass of the electron, the mass of the proton, the strength of the four forces, the initial expansion rate, etc. We know with great precision the values (in our universe) of all these parameters, but we have no idea why each has the specific value that it does. None of our theories can explain why the electron's mass is $9.109,382 \times 10^{-28}$ grams. Each parameter value could just as well have been some other number within a wide range, as far as we know. All our physical theories would work just as well with substantially different values. Perhaps there are other universes in which the parameter values are different.

However, there's a problem with other values. It seems that if these 20 parameters had even slightly different values, the universe would be a lifeless void. (Since we can't do the experiment of changing any of the parameters and seeing what would happen, this last statement is certainly speculative, but is consistent with the tone of this chapter.) Different parameter values would not be a problem for our physical theories, but would likely lead to a universe with no stars, no carbon, and no life.

So how did we get lucky enough to have all the right values in our universe? Enter Smolin's theory of Cosmological Natural Selection. As the name implies, Smolin introduces Darwin into cosmology.

Smolin asks, what if every black hole creates a new universe from its singularity? A new universe is created that expands into a new and different spacetime, entirely separate from its parent universe. Further, what if the new universe has parameter values that are similar, but not necessarily identical, to the values in its parent universe? This reminds us of a living organism inheriting its parents' DNA, perhaps with some mutations. Universes that make more black holes will have more offspring (they are "fitter") and their parameters (their "genes") will be passed on more effectively. Universes that make few or no black holes are history. "Good" parameters, those that result in many black holes, will proliferate far more than "bad" parameters. After many generations of universes, all the universes that remain are very likely to have good parameters because the others die out.

This is very Darwinian. Every creature born now is the product of countless generations of natural selection. Whether it is a lion or a lamb, its DNA has been extremely finely tuned for survival.

Therefore, Smolin says, it is no more surprising that our universe has good parameters than it is that humans have good DNA.

That's fine if you're a black hole. What about us carbon-based life forms? For a universe to make black holes, it must first make stars. Carbon, oxygen, and other heavy elements are very effective in radiating excess heat and helping gas clouds condense into stars. Universes will make more black holes if they make stars that produce these essential elements and disperse them by exploding in spectacular supernovae.

We get to enjoy the energy and the atoms that stars produce in the universe's quest to make the greatest number of black holes. Life is an inevitable result of universes striving to procreate.

If this is true, universes and life will go on forever.

I hope you find all this as exciting as I do.

Bibliography

COLLECTIONS OF CELESTIAL PHOTOGRAPHS

Astronomy Picture of the Day: http:// antwrp.gsfc.nasa.gov/apod/

Chandra Space Telescope gallery: http://chandra.harvard.edu/photo/

Frances, Peter (2005): Universe, the Definitive Visual Guide (DK Publishing, New York)

Hubble Space Telescope gallery: http://hubblesite.org/gallery/

Moore, Patrick (1998): Atlas of the Universe (Cambridge University Press, UK)

Ratcliffe, Martin (2007): State of the Universe 2007 (Praxis Publishing, New York)

Spitzer Space Telescope: http://www.spitzer.caltech.edu

Voit, Mark (2000): Hubble Space Telescope, New Views of the Universe (Harry N. Abrams, Inc. New York)

FOR A GENERAL AUDIENCE

Astronomy magazine.

Barrow, John (2002): The Constants of Nature (Vintage Books, New York)

Boehm, Gero von (2005): Who was Albert Einstein? (Assouline Publishing, New York)

Boslough, John (1985): Stephen Hawking's Universe (Avon Books, New York)

Cole, K.C. (2001): The Hole in the Universe (Harcourt, Orlando FL)

Crease, Robert P. (2004): The Prism and the Pendulum (Random House, New York)

Crease, Robert P. (1986): The Second Creation (Macmillan Publishing, New York)

Isaacson, Walter (2007): Einstein, His Life and Universe (Simon & Schuster, New York)

Kanipe, Jeff (2006): Chasing Hubble's Shadows (Hill and Wang, New York)

Meadows, A.J. (2007): The Future of the Universe (Springer-Verlag London, UK)

Melia, Fulvio (2003): The Black Hole at the Center of our Galaxy (Princeton University Press, Princeton, NJ)

Rees, Martin (1997): Before the Beginning (Perseus Books, Cambridge MA)

Scientific American magazine.

Seife, Charles (2003): Alpha & Omega (Penguin Group, New York)

Wheeler, John Archibald (1998): Geons, Black Holes & Quantum Foam (W. W. Norton & Company, New York)

Thorne, Kip (1994): Black Holes & Time Warps (W. W. Norton & Company, New York)

Tyson, Neil deGrasse and Goldsmith, Donald (2004): Origins (W.W. Norton & Company, New York)

FOR THOSE WHO ENJOY MATH AND PHYSICS

Dolling, Lisa (2003): The Tests of Time (Princeton University Press, Princeton, NJ)

Einstein, Albert (1961): Relativity (Three Rivers Press, New York)

Gell-Mann, Murray (1994): The Quark and the Jaguar (Owl Books, New York)

Greene, Brian (1999): The Elegant Universe (W. W. Norton & Company, New York)

Greene, Brian (2004): The Fabric of the Cosmos (Alfred A. Knopf, New York)

Guth, Alan (1997): The Inflationary Universe (Perseus Books, Cambridge, MA)

Hawking, Stephen (1988): A Brief History of Time (Bantam Books, New York)

Kaku, Michio (1994): Hyperspace (Anchor Books, New York)

Kaku, Michio (2005): Parallel Worlds (Anchor Books, New York)

Levin, Frank (2007): Calibrating the Cosmos (Springer Science+Business Media, New York)

Penrose, Roger (2005): The Road to Reality (Alfred A. Knopf, New York)

Penrose, Roger (1997): The Large, the Small and the Human Mind (Cambridge University Press, UK)

Randall, Lisa (2006): Warped Passages (Harper Perennial, New York)

Silk, Joseph (2001): The Big Bang (W. H. Freeman & Company, New York)

Smolin, Lee (1997): The Life of the Cosmos (Oxford University Press, New York)

Smolin, Lee (2006): The Trouble with Physics (Houghton Mifflin Books, New York)

Smolin, Lee (2001): Three Roads to Quantum Gravity (Perseus Books, New York)

Wheeler, John Archiblad and Kenneth Ford (1998): Geons, Black Holes and Quantum Foam (W. W. Norton & Company, New York)

Woit, Peter (2006): Not Even Wrong—The Failure of String Theory (Basic Books, Perseus Group, New York)

FOR THOSE WITH STRONG PHYSICS SKILLS

Allday, Jonathan (2002): Quarks, Leptons and the Big Bang (The Institute of Physics, London UK)

Duric, Neb (2004): Advanced Astrophysics (Cambridge University Press, UK)

Glendenning, Norman K. (2000): Compact Stars (Springer-Verlag, New York)

Hawking, Stephen and Penrose, Roger (1996): The Nature of Space and Time (Princeton University Press, Princeton, NJ)

Klapdor-Kleingrothaus, H. V. (2000): Particle Astrophysics (The Institute of Physics, London, UK)

Liddle, Andrew (2003): An Introduction to Modern Cosmology (John Wiley & Sons, West Sussex, UK)

Liddle, Andrew (2000): Cosmological Inflation and Large-Scale Structure (Cambridge University Press, UK)

Peacock, John A. (2000): Cosmological Physics (Cambridge University Press, UK)

Peebles, P. J. E. (1993): Principles of Physical Cosmology (Princeton University Press, NJ)

Phillips, A. C. (1994): The Physics of Stars (John Wiley & Sons, West Sussex, UK)

Raychaudhuri, A. K. (1992): General Relativity, Astrophysics, and Cosmology (Springer-Verlag, New York)

Weinberg, Steven (1992): Dreams of a Final Theory (Vantage Books, New York)

Wheeler, J. Craig (2007): Cosmic Catastrophes (Cambridge University Press, UK)

FOR GRADUATE PHYSICISTS

Misner, Thorne and Wheeler (1970) Gravitation (W. H. Freeman & Company, New York)

Rovelli, Carlo (2004) Quantum Gravity (Cambridge University Press, UK)

Weinberg, Steven (1970): Gravitation and Cosmology (John Wiley & Sons, West Sussex, UK)

Glossary of Terms

Abell: catalog of galaxy clusters by George Abell

absolute zero: the coldest possible temperature, 0 Kelvin or 0 K (–460 °F), at which there is no heat and all atomic motion stops

acceleration: a, the rate of change of velocity; a car going from 0 to 60 mph in 6 seconds has $a=10$ mph/second

accretion disk: a cloud of material swirling around a massive body, such as a black hole

action-at-a-distance: the purported ability of one object to affect another without direct contact, now discredited

amplitude: half the difference between a wave's maximum and minimum values

annihilate: to destroy completely leaving no material residue; antimatter and matter annihilate one another leaving only energy

anti-: prefix identifying an antimatter entity such as antiquark, antiproton, antielectron, etc.

Arp: catalog of "peculiar" galaxies by Halton Arp

atom: a component of matter consisting of a massive but minute nucleus surrounded by a diffuse cloud of electrons

AUI: American Universities Inc.

AURA: Association of Universities for Research in Astronomy

Big Bang: expansion of the universe from an original infinitesimal object of immense temperature

Big Bounce: cosmological scenario with a collapse prior to current expansion phase of the universe

Big Crunch: possible recollapse of the universe, now deemed unlikely

Big Rip: accelerating expansion of the universe; distant objects will disappear as they move away faster than the speed of light

black body spectrum: intensity of thermal radiation that varies in a specific way at different frequencies

black hole: a collapsed mass within a singularity of almost zero size surrounded by an event horizon at which the escape velocity equals the speed of light

blueshift: increase of frequency of light due to the source moving toward us, opposite of redshift

blue-white supergiant: a very massive, very hot, short-lived star

Bose-Einstein statistics: property of bosons to be gregarious.

bosons: a group of gregarious particles, carriers of nature's forces, preferring to coexist in a common state with others of the same type

brown dwarf: a proto-star without enough mass to sustain nuclear fusion and thereby become a true star

calculus: branch of mathematics for analyzing small changes (differential calculus) and combining these to determine global effects (integral calculus)

Caltech: abbreviation for the California Institute of Technology

Cepheid variables: stars whose brightness oscillates in consistent cycles and whose cycle duration is related to their maximum brightness

CFHT: Canada-France-Hawaii Telescope atop Mauna Kea, Hawaii

Chandrasekhar limit: maximum mass of a white dwarf

CMB: the first light, the radiation released when the universe first became transparent, now at a temperature of 2.725 K

CP-symmetry: the equality of particles and antiparticles

coherence: when multiple waves have the same frequency and fixed phase shifts

continuous: (1) of uniform composition without internal structure and not made of smaller parts; (2) without voids or abrupt changes

Cosmic Microwave Background radiation: see CMB

Cosmological Constant: Einstein's attempt to model a static universe

Cosmological Natural Selection: Lee Smolin's hypothesis that universes evolve to produce the greatest number of black holes

cosmology: the study of the universe as a whole entity

curvature: the bending of space or spacetime, as in Einstein's Theory of General Relativity

dark energy: an incompletely understood form of energy with repulsive gravity pushing the universe to expand at an ever faster rate

dark matter: an unknown form of matter exerting a gravitational force, not affected by the strong or electromagnetic forces

deuterium: an atom whose nucleus has one proton and one neutron; a rare isotope of hydrogen

diffraction: a wave spreading after passing through a small opening

discrete: made of individual, identifiable pieces

duality: two seemingly incompatible properties within a single entity

Einstein Field Equations: $G=8\pi T$, of General Relativity. T represents mass and energy, G represents curvature of space and time. Mass and energy tell space and time how to curve; space and time tell mass and energy how to move.

Einstein rings: light from distant sources focused into circular arcs by the gravity of intervening massive bodies

electromagnetic force: the force between charged or magnetic bodies; the force that hold atoms and molecules together

electron: an elementary particle with negative electric charge typically surrounding atomic nuclei

energy: the currency of existence. Its many forms include: potential, kinetic, mass, work, and heat. Energy is conserved—its total amount never changes; it can neither be created nor destroyed.

ether (luminiferous): purported medium through which light travels, no longer thought to exist

ESA: European Space Agency

ESO: European organization for astronomy, southern hemisphere

escape velocity: the speed required to escape from the gravitational field of a massive body

Euclidean: obeying Euclidean geometry, in particular having the interior angles of all triangles sum to 180 degrees

event: a location in four-dimensional spacetime

event horizon: the set of all points where the escape velocity from a black hole equals the speed of light

Fermi-Dirac statistics: property of fermions to be antisocial

fermions: a group of antisocial particles that are constituents of matter and do not share states with others of the same type

flat: having Euclidean geometry, used even when describing spaces with three or more dimensions

frequency: f, the number of full cycles per second of a wave

FRW equation: relates universe's expansion rate to its average energy density; a solution of Einstein's Field Equations in a homogenous, Euclidean universe

gamma ray: a photon whose energy is in the highest range, with wavelength less than 10^{-12} meters

General Relativity: Einstein's theory of gravity and the curvature of space and time

generation: elementary particles are grouped into three generations of increasing mass; does not imply some are descendants of others.

Goldilocks Zone: an optimal region for the existence of life

gluon: an exchange boson of the strong force

GPS: Global Positioning System providing precise position and velocity using satellites

gravitational lensing: focusing of light by massive bodies in accordance with Einstein's Theory of General Relativity

graviton: assumed exchange boson of gravity, not yet observed

gravity: described by Newton as a force attracting massive objects to one another; described by Einstein as the effect of spacetime curvature caused by all forms of energy

gravity waves: ripples in curved spacetime moving at the speed of light caused by the motion of massive bodies

habitable zone: where water may be liquid, typically a zone near a star

Hawking radiation: the light emitted near the event horizon of a black hole, reducing its mass, and eventually leading to its evaporation

heat: energy due to temperature, the kinetic energy of vibrating atoms

heavy water: a water molecule in which a normal hydrogen atom is replaced by deuterium, which is a heavier isotope of hydrogen

Hubble's Law: all distant galaxies are moving away from us at speeds that are proportional to their distances d: $v = Hd$

HUDF: Hubble space telescope Ultra-Deep Field image

homogeneous: the same everywhere

hyperspace: a space with more dimensions than normal; in the context of cosmology and General Relativity, a five- or more dimensional space within which our four-dimensional spacetime may exist

incoherent: waves with different frequencies or varying phase shifts

inertial frame: a reference system that moves with constant velocity. Earth's surface is generally accepted as an inertial frame. Newton's Laws and Einstein's Special Relativity apply only in inertial frames.

Inflation: a brief era of extraordinarily rapid expansion of the universe

infrared (IR): light with energy below visible light and above microwave, with wavelength in the range of 10^{-5} meters

interference: two or more waves combining

interference, constructive: waves combining with zero phase shift and reinforcing one another

interference, destructive: waves combining with phase shift of one-half wavelength and cancelling one another

interferometer: high-precision optical device to compare light travel times along two different paths

ion: an atom with a non-zero electric charge

isotope: an atom of the same element with the same number of protons but a different number of neutrons in its nucleus

invariant: having the same value in all reference frames

jet: a collimated stream of particles expelled from an energetic source, such as an accretion disk of a black hole

JPL: Jet Propulsion Laboratory, operated for NASA by Caltech

koan: an unstable particle formed from a quark and an antiquark

kinetic energy: the energy an object has due to its motion

lepton: an elementary particle unaffected by the strong force

light: electromagnetic radiation composed of oscillating electric and magnetic fields, and made of individual particles—photons

light echoes: light received after an initial flash due to the echo traveling a longer path

light-year: the distance light travels in one year, about 6 trillion miles

Local Group: the galaxy cluster containing our Milky Way, Andromeda, and over 40 smaller galaxies

lookback time: how long ago light that we see today was emitted, allowing us to observe the past

LQG: Loop Quantum Gravity, a theory combing Quantum Mechanics and General Relativity, now in development

LSP: Least-massive Super-symmetric Partner, possibly dark matter

luminosity: amount of radiated energy, such as by a star

M1-M110: objects known not to be comets catalogued by Messier

MACHO: Massive Compact Halo Object, possibly dark matter

macro-world: everything larger than a molecule, where Quantum Mechanics has negligible effects

Maxwell's equations: the equations of electromagnetism

mass: a measure of the amount of material in an object

matter: normal matter is made of protons, neutrons, and electrons

medium: the material waves travel through; air is a medium for sound

metric: equation for distance between points in a curved geometry

microwave: light with energy between infrared and radio wave, with wavelength in the range of 10^{-2} meters

micro-world: everything the size of a molecule or smaller, where the rules of Quantum Mechanics dominate

Milky Way: our galaxy, containing over 200 billion stars

Miracle Year: 1905, during which Einstein published five spectacular papers revolutionizing physics

molecule: two or more atoms bound together by sharing electrons

Msun: abbreviation for the mass of our Sun

muon: an elementary particle, a heavier version of electron

NRAO: U.S. National Radio Astronomy Observatory.

NOAO: U.S. National Optical Astronomy Observatory.

NASA: U.S. National Aeronautics and Space Administration, launches and operates most of the world's great space telescopes

natural units: a system of measurement units in which the speed of light c=1 and Newton's constant G=1

NGC: New General Catalog of galaxies by Dreyer and Herschel

nebula (nebulae): a cosmic cloud (clouds) of illuminated gas

neutral: having zero net electric charge

neutrino: an uncharged lepton, an elementary particle with extremely small mass and weak interactions with all types of particles

neutron: a particle with zero electric charge, made of 3 quarks, typically found only within an atomic nucleus

neutron star: a collapsed star with mid-range mass consisting primarily of neutrons

nuclear fission: splitting of a large atomic nucleus

nuclear fusion: small nuclei merging to make a larger one

nucleus: the core of an atom containing almost all its mass and energy

particle: one of a group of subatomic pieces of matter; particles of the same type are absolutely identical

particle, elementary: a particle not composed of other particles

particle-wave duality: the property of an entity having both particle and wave characteristics that seem incompatible

Periodic Table: chart of the elements arranged according to their chemical properties and the number of protons in their nuclei

phase shift: difference between positions of crests of two waves of same frequency

photoelectric effect: light incident on metal surface ejects electrons if its frequency is high enough

photon: a particle of light, exchange boson of electromagnetic force

Planck length: perhaps the smallest distance that exists in our universe, equal to 6×10^{-34} inches or 1.6×10^{-35} meters

Planck mass: possibly the total mass of the universe at the moment it came into existence, equal to 3×10^{-7} ounces or 2.2×10^{-5} grams

Planck time: possibly the smallest time interval that can exist, equal to 5×10^{-44} seconds

planetary nebula: a cloud of gas blown away by a dying star; it is made of star dust and has nothing to do with planets

plasma: a state of hot matter in which electrons separate from nuclei; plasma is opaque

potential energy: the energy an object has due to its distance from the center of a gravitational field

properly written: is defined in this book to mean written in terms of tensors in four-dimensional spacetime, thus making them valid in all reference frames and for all observers, regardless of their state of motion or position in gravitational fields

proton: a particle with positive electric charge, made of three quarks, component of atomic nuclei

proto-planet: a body that may develop into a planet

proto-star: a body that may develop into a star

pulsar: a neutron star with an immense magnetic field aligned askew to its axis of rotation

quantum: smallest possible unit, such as of energy; a penny is the quantum of U.S. currency

quantum fluctuation: a miniscule variation required by the Uncertainty Principle of Quantum Mechanics

quantum foam: breakdown of continuous spacetime at minute distances

Quantum Gravity: a hoped-for theory that will unite General Relativity and Quantum Mechanics

Quantum Mechanics: physical theory of the very small

quantum superposition: existing simultaneously in different states

quark: an elementary particle affected by the strong force, components of protons and neutrons; 6 types exist: up (u), down (d), charm (c), strange (s), top (t), and bottom (b)

quasar: a superluminous galaxy driven by a voracious black hole

qubit: a single storage element in a quantum computer

radio telescope: a telescope to image sources that emit radio waves

radio wave: light whose wavelength is more than 1 meter

redshift: reduction of frequency of light due to any of the following:
(1) light source moving away from us
(2) expansion of space
(3) light moving against the force of gravity

red giant: a greatly expanded star near end of life

red dwarf: a dim star with minimum mass to sustain nuclear fusion

resolution: ability to distinguish two slightly separated sources of light; higher resolution provides more detailed images

Schwarzchild metric: an important solution of Einstein's Field Equations for the spacetime curvature around a massive, round object

Schwarzchild radius: the radius of the event horizon of a symmetric, non-rotating, black hole

singularity: place where mass is concentrated in almost zero volume: center of a black hole, or the instant the universe came into existence

SOHO: Solar and Heliospheric Observatory satellite

solar wind: radiation and charged particles emitted by the Sun

spacetime: four-dimensional union of space and time

spacetime curvature: deviation from flat Euclidean geometry due to any form of energy, resulting in "gravity"

spaghettification: tidal forces stretching objects in the radial direction to a gravitating mass and squeezing them laterally

Special Relativity: Einstein's theory of space, time, and motion, excluding gravity and other forces

spectrum: collection of light frequencies an object emits or absorbs

star: a ball of gas so massive it sustains nuclear fusion in its core

standard candle: a distant source emitting a known amount of light

stellar wind: radiation and charged particles emitted by a star

strong force: force that holds together quarks and nuclei

supernova: explosion of a star releasing immense energy, and possibly creating a collapsed core: a neutron star or black hole

tensor: mathematical entity conforming to Special Relativity; tensor equations are valid in all references frames, even with gravity.

Thermodynamics: the physical theory of heat

thought experiment: a mental exercise focusing on key physical questions in an idealized situation

tic: defined here to be 10^{-30} seconds

time dilation: the slowing down of time that we perceive: (1) in a system with high relative velocity, due to Special Relativity; or (2) near a massive body, due to General Relativity

tidal force: differential forces of gravity that cause tides on Earth due to Moon and Sun, and spaghettification of objects near black holes

TLA: Three-Letter Acronym, a NASA specialty

ultraviolet (UV): light with energy above that of visible light and below x-rays, with wavelength in the range of 10^{-8} meters

universe: defined to be all we can observe or be influenced by, in any conceivable manner

universe, expansion: increase in the distance between any two distant points in the universe

variable stars: stars whose brightness cycles periodically

velocity: *v*, an object's speed and direction, e.g. 60 mph due north

velocity, absolute: discredited notion that velocities measured in a special, "absolute" frame of reference have greater importance than those measured in other frames

velocity, relative: motion measured with respect to something else, e.g. Earth moves 70,000 mph relative to the solar system

virtual particles: particle-antiparticle pairs spontaneously created, existing briefly, and vanishing as allowed by the Uncertainty Principle

wave: an assembly of many small objects moving together in an oscillatory manner

wavelength: *w*, the distance between wave crests

wave packet: a partially localized combination of waves of different frequencies

weak force: the force enabling radioactive decay and allowing more matter than antimatter to develop in first 1 second after Big Bang

white dwarf: a collapsed star with mass < 1.4 *Msun*

WIMP: Weakly Interacting Massive Particle, possibly dark matter

WMAP: NASA's Wilkinson Microwave Anisotropy Probe satellite

x-ray: a photon with energy between gamma rays and ultraviolet, with wavelength in the range of 10^{-10} meters

yellow dwarf: a star like our Sun, sustained by hydrogen fusion, with mid-range mass and temperature

Index

The most important references are shown in **bold** type.

acceleration, **51**, 65, 76, 99, 152–154, 160, 188–191
accretion disk, **193**–195, 201, 206
Allday, Jonathan, 106
Andromeda Galaxy, 164, **204**–207, 229, 236, 264
antimatter, **46**–49, 117, 129-130, 196, 261, 282–283
antiparticles, see antimatter
Arecibo Radio Telescope, 198, **226**–228
Aristotle, 82–84
atom, 11, 54, 68–69, 134–135, 143, 214, 229, 283–285
    composition and structure, 11, **34**–36, **112**–113, 120–122, 131
    debate about existence, 16–19, 28–29, **32**–33, 107
Big Bang, 13, 52, 134–135, 143, 201, 240, 246–249, **274**–291
Big Bounce, 291
Big Crunch, 243
Big Rip, 260
black hole, 11–12, 52, 71, 142, 152, 163–166, 172, 180, **182**–197, 208, 216–217, 223, 264–265, 277–278, 291–293
    see accretion disk, event horizon, jet, and singularity
blueshift, 11, 101, 104, **230**–232
blue-white supergiant stars, 137
Bohr, Niels, 45, 103, 109, 112–113, **115**–118, 124–125
Bojowald, Martin, 291
Boltzmann, Ludwig, 17–18, 28
Born, Max, 160–161
Bose, Satyendra Nath, **52**, 108, 131

boson, 39, 44, 46, **51**–56, 131, 256

de Broglie, Louis, 111–112

Brownian Motion, **28**–29, 32–33, 107

Bullet Cluster, 256

Caltech, 30, 53, 186, 198, 200, **207**, 208, 255

Casimir effect, 261, **263**

Cepheid variable stars, 236

Chandrasekhar, Subrahmanyan, 176–177

Chandra X-ray Space Telescope, 8, 177, 195, 198, **200**–201, 294

coherence, **123**–126, 131

Comte, August 113

cosmic microwave background radiation, (CMB), 86, 101, 200,
    228, 241, **246**–252, 254–255, 260, 276, 280, 283–284

cosmological constant, **235**, 262

Cosmological Natural Selection, 292–293

cosmology, 13, 198, 199, 229–230, 246–263, **274**–293

Crab Nebula, **178**, 201

Crease, Robert, 40, 106

curvature of space or spacetime, 58, 106, 147–151,
    155–165, 178, 184, 187–188, 251–252, 277, 279

dark energy, 243, 252, **260**–263, 275, 289

dark matter, 252, **254**–258, 260, 262, 275

Democritus, **17**, 33, 38

diffraction, **76**–78, 126

Dirac, Paul, 160–161

dwarf stars,
    brown, 137
    red, 137
    white, **170**–174, 176–177
    yellow, 137

Eagle Nebula, 135

Earth, 30, 36, 51, 56–57, 72, 85–88, 139–141, 143–148,
    153–159, 164, 184, 185, 187, 200, 204, 207–209, 212,
    224, 244, **264**–268, 270–273, 285–286

Eddington, Sir Arthur, 106, **150**–151, 160–161, 177

Einstein, Albert, 10–13, **20**–28, 45, 132, 176, 292
    achievements, 104–**107**, 237
    atoms, 28, **32**-33, 107
    Bohr, debate with, **115**–119, 124
    Bose-Einstein statistics, **52**, 107, 108, 131, 181
    diffusion equation for Brownian motion, 32–33
    E=mc², 48, 60, **62**–66, 107
    Einstein Cross, 166–**167**, 223
    Einstein rings, **167**, 209–210
    Equivalence Principle of General Relativity, 153–155
    Field Equations of General Relativity, **160**–165, 229, 235,
        262–263, 280–281
    General Relativity, 12–13, 52, 90, 106, 107, 117, **149**–169, 179–181,
        187, 215, 216, 229, 234–235, 237, 251, 261–263, 277–281
    gravitational lensing, 166–168
    gravitational redshift, 169
    gravity, 51, 58, **147**–159
    gravity waves, 169, **179**–181
    lasers, 107, **131**
    light, 79–80, **100**–102, 107
    marriage to Maleva Maric, 23–**24**, 245
    Miracle Year, **104**-105
    Nobel Prize, 80, **103**, 245
    particle-wave duality, 74–75, **80**, 107, 108, 111
    photoelectric effect, **79**, 107
    precession of Mercury, 168–169
    Quantum Mechanics, **108**, 110–111, 115–119, 124, 131
    quotes, 21, 95, 117, 118, 124, 132, 161, 202, 237, 292, **320**
    Special Relativity, 22, **90**–99, 102, 107, 244, 280
    starlight bending, **150**–151, 166–168
    youth, 20–22
electromagnetism, **50**–57, 87, 91, 100–102, 244, 254, 263, 268
electron, 11, **34**–36, 38–41, 46–49, 53, 56, 60, 68, 79, 110,
        112–113, 117, 120–121, 128–131, 161, 176–178, 283, 292
electronics, 12, 67, **122**

energy, 18, 34, 47–49, **60**–65, 79, 94, 100–102, 107, 109–116, 120–122, 129–131, 148–149, 160, 184–185, 217, 228, 250–252, 273
generation, 11, **66**-73
see dark energy
ether, 26, **87**–89, 91–92
event horizon, 164, **183**–190, 192, 196–197, 216–217
Faraday, Michael, 54–55
Fermi, Enrico, 39, **43**, 45, 201, 225
fermion, **39**–44, 131
Feynman, Richard, 30–**31**, 53, 109, 129, 259, 269
frequency, **75**, 79, 100–101, 110, 112–115, 131, 169, 230–231, 248
Friedman, Alexander, 235, 281
FRW equation, 280–281
galaxy, 57–59, 136, 164, 166–168, 192–195, **204**–213, 223, 228–229, 233, 235–244, 250, 255–258, 280, 283
Galilei, Galileo, 11, 50, **82**–84, 144, 149, 154, 199–200
General Relativity, see Einstein
Goldilocks Zone, 139, **265**, 267
gluon, 55
Global Positioning System (GPS), 13, **164**
gravitational lensing, 166–168
graviton, 46, **52**, 94, 102
gravity, 50–53, **57**–60, 106–107, 117, 135–142, 144–161, 164–170, 172–173, 177–178, 187–191, 234–235, 243, 254–258, 260–263, 267, 280
waves, 169, **179**–181
habitable, 72, 134, **140**, 267, 290–293
Hawking, Stephen, 288–289
radiation, **196**–197, 216–217
heat, **17**–18, 25, 56, 61–65, 66–67, 110, 135, 248–249, 266, 280, 293
heavy water, 71
Heisenberg & Uncertainty Principle, 45, **114**–119, 129–130, 196, 289
Hoyle, Fred, 276–277
Hubble, Edwin, 13, 198–199, **236**–240
Hubble's Law, 239–241
Hubble Space Telescope, 8, 13, 192, **198**–201, 209–211, 223, 242

Huchra, John, 167
Hulse, Russel, 178–180
hyperspace, **149**, 278
inertial frame, **86**–88, 92, 95–99, 101, 152
Inflation, 274–276, **278**–280, 289–291
    Eternal, 290
interference, **75**–78, 113–114, 123, 126–129
    interferometer, 88–89
jets, 170–171, **193**–194, 201
Jet Propulsion Lab, 200
Keck, twin telescopes, 199
Kelvin, Lord, 25–26
    temperature scale, 25, **134**, 217
Least-massive Super-symmetric Partner (LSP), 256
Leavitt, Henrietta, 13, **236**–237
lepton, 39–42
light, 22, 54–55, 78–80, **100**–102, 106–107, 110–111, 149–151,
        166–169, 186-187, 190, 223, 230–233, 246–252
    echoes, 191
    ether, 26, **87**–89, 91–92
    photon, **46**–49, 51–53, 79, 94–97, 100–102, 107–108,
        111–113, 131, 196, 246, 252, 263, 283
    spectrum, **101**–102, 112, 174, 200–201, 246, 248
    speed of (c), 22, 55, 63, **87**–96, 102, 201, 218–219, 222–224, 280
light-year, 57, 244
Linde, Andre, 290
Local Group, **205**, 244, 262, 264
lookback time, **242**–243, 317
Loop Quantum Gravity, 291
low energy neutron reactions, (LENR), 73
MACHO, 256
Maiman, Theodore, 131
Manhattan Project, **45**, 259
mass, 11, 34, 41–44, 46–47, **60**–67, 92–94, 107, 135–137,
        146–149, 158–160, 163–166, 171–172, 182–184, 255

matter, 11, 16–18, **34**–44, 46–49, 57, 60, 131, 176, 183, 194, 250–252, 258, 260–263, 282–283, 289

Maxwell, James Clerk, 54–55
    equations, 78, 87–88, 91–92

medium of a wave, 76, see ether

metric of spacetime, 165

Michelson, Albert, 88–89

Milky Way, 57, 86, 136, **204**–207, 228–229, 235–236, 244, 264–265

Minkowski, Herman, 22, **97**

Miracle Year, **104**–105

Monaldi, Daniela, 40

Moon, 60, 124, 150, 199–200
    creation of, 266
    effect on Earth, 187, 266
    orbit, 146–147

multiverse, 289

muon, **40**–41, 253

NASA, 8, 12, 136, 176, 185, 190, 192–193, **198**–201, 207, 249м 272м 274

natural units, **159**, 190, 255

nebula, 58, **135**, 170–171, 236

neutrino, **40**–44, 102, 117, 174, 177, 252

neutron, 11, 35–36, **39**–41, 47–49, 56, 131, 177–178, 282–284

neutron star, 12, 172, **176**-182, 201

nuclear fission, 45, 66, **68**–69, 72–73

nuclear fusion, 54, **68**–71, 135, 137–142, 172–174

nucleus, 11, **34**–36, 54, 68–69, 112–113, 119, 121, 142, 176, 178, 284

Newton, Sir Isaac, 11, 18, 50, 78, 86–87, 90, 94, 99, 110, 115, 263
    gravity, **144**–151, 168–169, 191, 234, 262, 280

particle, elementary, 11, **38**–49, 71, 115, 131, 135

particle-wave duality, **80**, 107–108, 111

Penrose, Sir Roger, 71

Penzias, Arno, 247–248

Periodic Table, 35, 131

Perrin, Jean Baptiste, 33

Pestalozzi, Johann, 22

Piccioni, Oreste, 40, **42**, 48, 253, 282
phase shift, **122**–123, 126–127
photoelectric effect, **79**, 107
photon, see light
Planck, Max, 105, 107, **110**–111, 235
    black body spectrum, 246, **248**
    constant, 79, 100, **110**, 115, 129
    length, 182–183, 214–**216**, 276, 278
    mass, 217
    time, 216
plasma, **194**, 249, 257
Poincaré, Henri, 26–27, 91
proton, 11, 35–36, **39**–41, 47–49, 54, 56, 131, 177–178, 214, 282–284
proto-planet, 266
proto-star, 137
pulsar, 178–180
quantization, 75–76, **109**–113, 248, 277
    of space, 182–183, **278**, 291
Quantum
    computers, 125–126
    Field Theory, 51–52
    fluctuations, 275, **280**, 289, 290
    foam, 278
    Gravity, 261, 274, 276, 278, **291**
    Mechanics, 12, 38–39, 53, **108**–131, 215, 261, 275, 277–278
quark, 11, **38**–42, 47, 54, 282–283
quasar, **166**–167, 223, 265
radio telescope, 195, **198**, 201, 226–228
red giant stars, **137**, 141, 171
redshift, 13, 169, **230**–233, 237–239, 246, **317**
resolution, 199–200
Relativity, 11–13, 22, **83**–99, 103, 107, 120, 153, 244, 275, 280
    see Galileo, Einstein, Special Relativity and General Relativity
Rovelli, Carlo, 278, 291

Rubin, Vera, 255

Schalow, Arthur, 131

Schroedinger, Erwin and his Cat, 124–126

Schwarzchild , Karl, 163

    metric, 164–165

    radius, 165

singularity, **182**–188, 293

spacetime, **97**–98, 106, 147–149, 165, 169, 179, 278–279, 281, 288–293

    see curvature of

spaghettification, 187–188

Smolin, Lee, 291–293

Spitzer Space Telescope, 8, 198, 200–201

standard candle, 175

starlight, 13, 106, **112**, 118, 150–151, 163, 166–168, 230–233

stars, 12–13, 64, 98, 112–113, **134**–143, 195, 204–206, 211, 229, 258,

    birth of, 135–139

    death of, 141, 171–172

    enable life, 12, 54, 69, **134**–135, 143, 284–285, 293

    lifetime of, 139–141

    luminosity of, 139–140

    mass range, 137

    neutron star, 172, **176**–182, 201

    star types, 137, see also dwarf and giant

    variable, 236–237

strong force, 11, 39–41, 50–**54**, 254, 282

Sun, 36, 51, 56–57, 64–65, 72, 86, 106, 136–143, 146–151, 163–164, 168–169,

    171–172, 188–191, 204–206, 212–213, 229, **265**–268, 273, 284–285

supernovae, 12, 136, **141**, 206, 293

    core collapse, 177

    SN1987a, **141**–142, 175, 177

    Type Ia, **173**–174, 238, 260

Supersymmetry, 256

tensors, **160,** 277

Taylor, Joseph, 178–180

Thorne, Kip, 186
time
    dilation, **96**, 98, 164
    imaginary, 288–289
    views of Newton vs. Einstein, 94–95
Thermodynamics, **17**–18, 110
Thomson, J. J., 160–161
thought experiments, 23, **95**, 116, 124
tidal forces, **187**–188, 196
Townes, Charles, 131
twin paradox, 98
two-slit experiment, 126–129
universe, 10–13, 134, 147–149, 202, **203**–293
    age, 216, 250–251
    contents, 36, 87, 204–211, 217, 226–229
    definition, 218
    evolution, 49, 223, 274–287
    expansion, 234–244, 260–262
    size, 211–215
velocity, 50–51, 93–96, 99, **146**
    absolute, **84**–88, 91, 96
    relative, **84**–86, 91–92
    escape, **184**–186, 243
Very Large Array (**VLA**), 198
virtual particles, 51, **129**–130, 196–197, 215, 260–263, 289
Wheeler, John Archibald, 160, 186, 278
Wilson, Robert, 247–248
waves, 22, 55, **74**–80, 87, 100–101, 112–115, 122–123, 126–129
    wavelength, 74–75, 100–101, 111–**115**, 169, 183, 216, 230-233
    wave packet, 113–115
weak force, 11, 50, **52**–53, 56, 256, 266, 282–283
Widom-Larsen theory of LENR, 73
WIMP, 256
WMAP, 8, 198, 200, **249**–251
Zwicky, Fred, 255

Table of Redshift vs. Time

For light observed with redshift z, table lists Lookback Time (how long ago this light was emitted) and Age of the Universe (how long after the Big Bang it was emitted). Table is based on FRW expansion dominated by matter and dark energy. See chapters 32 and 39 for further discussion.

Redshift z	Lookback Time in Billions of Years	Age of Universe in Billions of Years
0	0	13.7 = now
0.01	0.14	13.6
0.02	0.27	13.4
0.03	0.41	13.3
0.06	0.8	12.9
0.10	1.3	12.4
0.15	1.9	11.8
0.25	2.9	10.8
0.5	5.0	8.7
0.7	6.3	7.4
1.0	7.8	5.9
1.5	9.3	4.4
2.5	11.0	2.7
6	12.7	1.0
10	13.2	0.5
30	13.6	0.1
200	~13.7	0.01
1091	~13.7	0.000,38

List of Symbols

γ	*gamma*: symbol for photon
ν	*nu*: symbol for neutrinos; there are three types
Λ	*lambda*: Einstein's cosmological constant
a	acceleration, the rate of change of velocity
A	wave amplitude
c	speed of light in empty space: 671 million mph
d	distance
E	energy
$\mathcal{E}$	electric field
f	frequency, number of cycles per second of a wave
g	Einstein's relativistic factor in time dilation, etc.
G	Einstein curvature tensor in $G=8\pi T$
h	Planck's constant
H	Hubble expansion rate that varies over time
H_0	Hubble expansion rate now, at $z=0$
Hz	Hertz: the number of cycles per second
K	degrees Kelvin: 0 K is absolute zero temperature
m	mass, M is also used for the mass of a very large body
$\mathcal{M}$	magnetic field
Msun	mass of our Sun
r	radius
t	time
T	energy tensor in $G=8\pi T$
v	velocity: speed and direction, e.g. 60 mph due north
w	wavelength: distance between wave crests
z	redshift of light from stars or other cosmic sources

Visit Dr. Robert Piccioni's website:

www.guidetothecosmos.com

for:

- Best Discounts on Robert's Books, CDs, and DVDs

- Information about Live Presentations

- Videos of Robert's Lectures

- Robert's Radio Shows, including:
  "Protecting Earth from Asteroids" with Astronaut Rusty Schweikart
  "Energy and the Environment" with Dr. Reese Halter
  "Observatories Ancient and Modern" with Dr. Edwin Krupp
  "We Are Stardust" with Dr. Jerome Clifford
  "Green Energy That's 'CLENR' " with Lewis Larsen
  "World's Largest Telescopes" with Dr. Jonathan Romney
  "Andromeda Cometh" with Ron Chegwidden
  "$E=mc^2$ and Smarter Energy"
  "Talking to Deep Space" with Gene Burke of JPL
  "Einstein, Light, and the Odd Nobel Prize"
  "Alicia's Summer Job on Jupiter" with Alicia Andro
  "How our Moon was Made" with Dr. Dana Mackenzie
  "Photography That's Out Of This World" with Dr. Rob Gendler
  "Hey, Einstein, What Time Is It?"
  "Hunting Supernovae in Your Spare Time" with Tim Puckett
  "Imagine Our World Without Einstein"
  "World's Youngest Supernova Discoverer" with Caroline Moore
  "Henrietta Leavitt: First Famous Woman Astronomer"
  "Quantum Uncertainty and Reality"
  "Telescopes are Time Machines"
  "Dark Matter and the Bullet Cluster" with Dr. Marusa Bradac

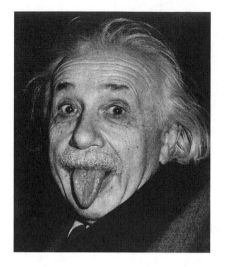

MORE EINSTEIN QUOTES

"The difference between stupidity and genius
is that genius has its limits."

"No amount of experimentation can ever prove me right;
a single experiment can prove me wrong."

"It should be possible to explain the laws of physics to a barmaid."

"When you are courting a nice girl an hour seems like a second.
When you sit on a red-hot cinder a second seems like an hour.
That's relativity."

"I know not with what weapons World War III will be fought,
but World War IV will be fought with sticks and stones."